Current Topics in Microbiology and Immunology

168

Editors

Hepadnaviruses

Molecular Biology and Pathogenesis

Edited by
W. S. Mason and C. Seeger

With 21 Figures

Springer-Verlag
Berlin Heidelberg NewYork
London Paris Tokyo
Hong Kong Barcelona
Budapest

William S. Mason
Christoph Seeger
Fox Chase Cancer Center
7701 Burholme Ave.
Philadelphia, PA 19111
USA

Cover design by courtesy of Dr. Allison Jilbert, Institute of Medical and Veterinary Science, Adelaide, South Australia.

ISBN-13:978-3-642-76017-4 e-ISBN-13:978-3-642-76015-0
DOI: 10.1007/978-3-642-76015-0

Softcover reprint of the hardcover 1st edition 1991
Library of Congress Catalog Card Number 15-12910

Typesetting: Thomson Press (India) Ltd, New Delhi;

23/3020-543210—Printed on acid-free paper.

Preface

The discovery of animal models and the possibility of expressing virus in tissue culture cells have catapulted hepadnavirus research from auspicious beginnings more than two decades ago into a dynamic field that has in a short time uncovered the major problems pertinent to the viral replication cycle. However, human hepatitis B virus (HBV) still infects more than 200 million people worldwide and remains one of the most severe human pathogens, despite the availability of efficient vaccines to prevent primary infection. This is a consequence both of the ability of the virus to cause chronic, life-long infections and of the failure of conventional therapies to block virus production and effect a permanent cure of the infected liver. The goal of curing a chronic carrier may, however, be accomplished in the near future as the molecular principles of hepadnavirus replication continue to unfold. It is the purpose of this volume to summarize the results of the latest efforts by investigators in this field and to provide perspectives, emerging from today's knowledge, for the years to come.

The book commences with a chapter by Israel and London describing the consequences of HBV infection on anatomy and function of the liver. One of the major characteristics of hepadnaviruses is their marked hepatoropism. This property is, in fact, believed to be imprinted at least on one level by sequence elements encoding viral promoters and enhancers. Schaller and Fischer discuss how viral gene expression is regulated and how the mechanism of gene expression controls virus production in the infected hepatocyte. The next two chapters deal with the mechanism of viral replication in the liver. Investigations of the mechanism of viral DNA synthesis have long been hampered by the failure to propagate hepadnaviruses in cultured cells. The development of tissue culture systems for the expression of infectious virus has energized a genetic and functional analysis of the hepadnavirus genome. The results from these investigations and current models explaining how hepadnaviruses synthesize DNA from RNA are summarized in the chapter by Seeger et al. Biogenesis of the viral particle is the subject of the next chapter by Ganem, who describes the strategy employed for the synthesis of the viral proteins and discusses the current knowledge of the signals controlling assembly of infectious virus. The following two chapters are devoted to the most severe consequence of hepatitis B virus infection, the development of hepatocellular carcinoma. To date, much of our understanding of the process of hepatocarcinogenesis has been based upon analogies to other systems. Accordingly, Chisari discusses how transgenic mouse models have been used to test some novel ideas about the causes of liver cancer in an infected host. Rogler then describes what is known about the role of hepadnaviruses in hepatocarcinogenesis in their indigenous hosts and discusses current molecular models of liver cancer ontogeny in this setting. An important clue for unraveling the parameters commanding viral gene expression may emerge from the molecular analysis of hepatitis delta virus (HDV), discussed by Taylor. HDV replication, while itself dependent on one or several HBV products, suppresses replication of hepadnaviruses by an as yet unknown process. The closing two chapters of this volume provide a perspective of the long-term goals of hepadnavirus research; the first, by Marion, considers central questions about the development of antiviral therapies for chronic infection, and the second, by Horwich, the possibility of using hepadnaviruses as vectors for liver-directed gene therapy in inherited diseases.

William S. Mason
Christoph Seeger

List of Contents

List of Contributors

(You will find the authors addresses at the beginning of their respective contributions)

Liver Structure, Function, and Anatomy: Effects of Hepatitis B Virus

J. ISRAEL[1] and W. T. LONDON[2]

Hepatitis B virus (HBV) is a common human pathogen that causes extensive morbidity and mortality in human populations throughout the world. Although the virus has been identified in extrahepatic tissues, such as vascular endothelium, bile duct epithelium, bone marrow, and peripheral blood lymphocytes (BLUM et al. 1983), the liver is the major and most important site of infection. It is the disruption of the normal anatomy and functions of the liver that leads to a set of clinically and pathologically defined acute and chronic diseases, including acute and chronic hepatitis and cirrhosis. Chronic infection with HBV is also associated with the development of hepatocellular carcinoma, particularly in the setting of a cirrhotic liver. None of these diseases is specific to HBV infection;

[1] Gastrointestinal Section, Department of Medicine, University of Pennsylvania, Philadelphia, PA 19104, USA
[2] Division of Population Sciences, Fox Chase Cancer Center, 7701 Burholme Avenue, Philadelphia, PA 19111, USA

Current Topics in Microbiology and Immunology, Vol. 168

each can be caused by other infectious and hepatotoxic agents. What is distinctive about HBV is that it can cause all of these diseases as well as no apparent disease. It is this variability in outcome that is both the least understood and the most fascinating to virologists and clinicians.

The purpose of this introductory chapter is to describe the normal structure and functions of the liver and the array of alterations that can be induced by acute and chronic infections with HBV.

1 Liver Anatomy

1.1 Gross Anatomy

The liver is a large solid organ, weighing 1200–1500 g in adults, located in the right upper quadrant of the abdomen and partially protected by the ribs (SHERLOCK 1989). It is composed of a large right lobe, which is subdivided into quadrate and caudate lobes, and a smaller left lobe (Fig. 1). While the two main lobes are distinguished by ligaments, the entire organ is a contiguous parenchymal mass and the lobes are not delimited by any fibrous septum that traverses the liver substance. The vascular and biliary systems of the different lobes are interconnected (RAPPAPORT 1987).

Blood is supplied to the liver in a unique fashion that is well suited to several aspects of its functions, particularly those involving nutrient metabolism and drug detoxification. Well-oxygenated arterial blood is supplied by the hepatic artery, while venous blood is carried to the liver from the intestines and spleen by the portal vein. The portal system brings absorbed carbohydrates, fats, and proteins, as well as drugs and other toxins directly from the gut to the liver. Both the portal vein and hepatic artery enter the liver at the porta hepatis where they divide to form separate branches to the left and right hepatic lobes. Blood is carried from the liver by two or more hepatic veins that open into the inferior vena

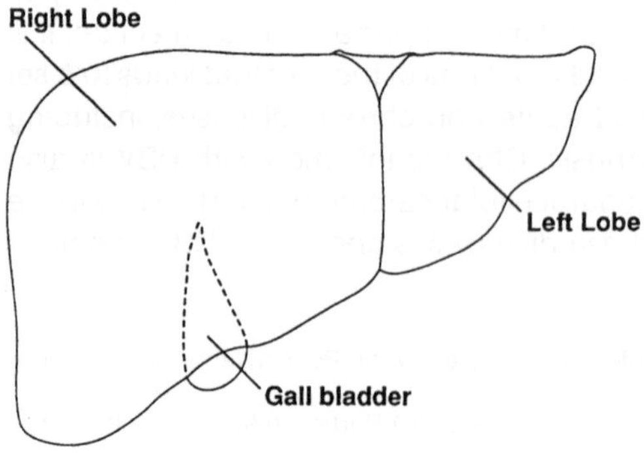

Fig. 1. Anterior view of the human liver. The liver fits snugly under the dome of the diaphragm and is protected anteriorly by the rib cage. With the ribs removed, only the right and left lobes of the liver and the inferior pole of the gall bladder are visible. The quadrate and caudate lobes are subdivisions of the right lobe located on the posterior (caudate) or inferior (quadrate) aspects of the liver

cava near where it enters the heart. Lymph vessels from the liver terminate in glands near the porta hepatis, along the inferior vena cava in the thorax and mediastinum (SHERLOCK 1989).

The biliary system, which originates in the liver, supplies bile to the small intestine where it aids in fat absorption. Bile is secreted by hepatocytes into the bile canaliculi, transported and modified by the intrahepatic ductules to exit the liver through the common hepatic duct (ERLINGER 1987). From the common hepatic duct, bile flows into the common bile duct. The cystic duct is a branch off the common bile duct that connects it with the gall bladder. The gall bladder, while not technically part of the liver, is anatomically and functionally related to it. Located on the under surface of the right lobe of the liver, the gall bladder is a pear-shaped organ that has a capacity to hold 50 ml of fluid. It stores hepatic bile until a hormonal signal, in response to food in the small intestine, induces contraction, causing bile to course back through the cystic duct into the common bile duct. From there, bile continues distally and empties into the intestine (duodenum) through the ampulla of Vater, a channel formed by the confluence of the common bile duct and the pancreatic duct.

Shortly after food enters the small intestine, bile and pancreatic juice are secreted into it. These secretions act on the nutrients to begin the process of digestion. The major constituents of bile are bile acids which are synthesized in hepatocytes and secreted into bile ductules. After aiding in the digestive process, which occurs throughout much of the small intestine, the bile acids are actively reabsorbed in the terminal ileum (the very end of the small intestine). After absorption, the bile acids enter the portal system and are efficiently removed by the liver where they are promptly secreted into bile once again. The entire anatomic route taken by these important transport molecules is called the enterohepatic circulation (HOFFMAN 1989).

1.2 Microscopic Anatomy

The liver has been divided into structural subunits, called lobules, based primarily on its two-dimensional appearance in the light microscope. First described by KIERNAN in 1833, a lobule consists of a polyhedral section of liver tissue which appears hexagonal on cross-section. It is composed of a central vein (a tributary of the hepatic vein) and, at the periphery, portal tracts (triads) each of which contains a bile duct, portal vein tributary, and a branch of the hepatic artery (Fig. 2). The central hepatic vein and the portal triads are connected by cords of liver cells and sinusoids (contiguous spaces between the liver cells). In three dimensions, the cords of liver cells are really sheets of hepatocytes, one cell in thickness (ELIAS 1949; GRISHAM 1976). Under normal circumstances the portal vein has a higher pressure than the central venous system so that blood flows from the portal system into and through the sinusoids towards the central hepatic vein (SHERLOCK 1989). The arterial system follows the same path as the portal system resulting in a mixing of arterial and portal venous

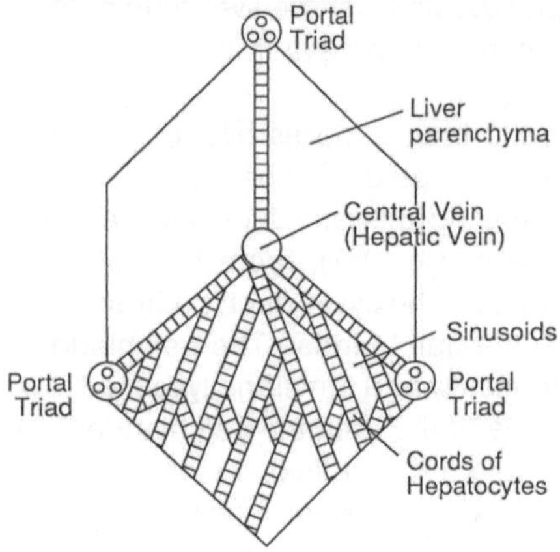

Fig. 2. Schematic representation of a hepatic lobule based on two-dimensional light microscopy. The central vein in the center of the hexagonal lobule is a tributary of the hepatic venous system. Each portal triad contains a tributary of the portal vein, a branch of the hepatic artery, and a bile duct. The sinusoids are blood-filled spaces between cords of hepatocytes. Both portal and arterial blood flows into the sinusoids and from there into the central veins (SHERLOCK 1989)

blood in the sinusoids. Sphincters in each system regulate flow into individual sinusoids.

Since their original description, the in vivo existence of hepatic lobules has been questioned (AREY 1932; BRITTAIN et al. 1964; OPIE 1944; MALL 1906; RAPPAPORT 1953; RAPPAPORT et al. 1954). The concept of a microscopic unit of liver tissue filling a hexagonal space around a central vein breaks down when one attempts to integrate these individual units into an overall structure for the liver. In addition, the concept of a hexagonal lobule is not compatible with observations of the hepatic microcirculation in situ (RAPPAPORT 1987).

Based on studies of the microcirculation, the liver has been divided into microscopic units called acini (RAPPAPORT 1976). Each acinus is an irregular mass of parenchymal tissue centered on a portal triad which again consists of a portal vein, bile duct, and hepatic artery. The acinus lies between and drains into two hepatic (central) veins (Fig. 3). It has no surrounding capsule and by two-dimensional light microscopy appears to occupy part of two adjacent hepatic lobules. The liver parenchyma immediately surrounding the portal triad is referred to as zone 1. Zone 3 lies at the periphery of the acinus, near the hepatic veins. This zone receives the poorest supply of oxygenated blood and is the area that suffers the most injury from anoxic, toxic, or viral insults (SHERLOCK 1989). Zone 2 is located between zones 1 and 3 and is intermediate, as well, in terms of blood supply.

In addition to hepatocytes, the liver is composed of endothelial cells, Kupffer cells, and bile duct epithelial cells. Sixty percent of the cells in the liver are hepatocytes. These are polygonal cells approximately 30 µm in diameter. In relation to their environment, hepatocytes have three surfaces, but no basement membrane. One surface of the hepatocyte faces the sinusoid and the space of Disse, while a second surface borders the bile cannaliculus, and the third touches neighboring liver cells. The plasma membrane of hepatocytes is reinforced by microfilaments forming a supportive cytoskeleton. The surface of hepatic plates is sealed by junctional complexes between individual hepatocytes, including tight

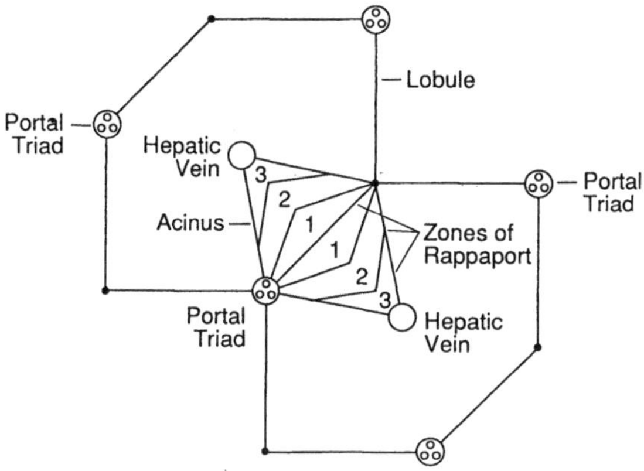

Fig. 3. Schematic representation of the hepatic acinus as proposed by RAPPAPORT (1976). The hepatic acinus is centered on a portal triad and lies between two hepatic (central) veins. The parenchymal tissue closest to the portal triads was designated by Rappaport as zone 1. It receives the best-oxygenated blood and is the area most resistant to liver injury. Zone 3 is closest to the hepatic veins, receives the poorest supply of oxygenated blood, and is most susceptible to liver injury (RAPPAPORT 1976; SHERLOCK 1989). Zone 2 is intermediate in oxygen supply and intermediate in susceptibility to liver injury

junctions, gap junctions, and desmosomes. In a healthy liver an occasional mitosis can be seen. The lifespan of a liver cell in experimental animals is about 150 days.

The sinusoids are hepatic capillaries lined with endothelial cells. Kupffer cells are phagocytic cells that form part of the sinusoids and comprise a major part of the reticuloendothelial system. The space of Disse is a tissue space between the hepatocytes and the sinusoidal lining cells. Bile canaliculi form as grooves on the contact surfaces of liver cells. The canalicular network drains near the portal vein into thin-walled bile ductules that are lined with cuboidal epithelial cells. The ductules terminate in the bile ducts (SHERLOCK 1989).

2 Major Hepatic Functions

Normal liver function is essential for human life. While technology has made it possible for individuals without kidneys to continue living through hemo- or peritoneal dialysis, no such techniques exist to compensate for the loss of normal hepatic function. The liver is essential for carbohydrate, protein, and lipid metabolism. Furthermore, it plays an important part in bilirubin, drug and toxin, vitamin and metal metabolism (Table 1). Severe liver disease, such as cirrhosis from a variety of causes, results in predictable symptoms from derangements in these metabolic pathways. While no artificial means exists to assume the essential metabolic activities of the liver, orthotopic liver transplantation in selected

Table 1. Major metabolic functions of the liver

1. Carbohydrate	Homeostasis of blood glucose levels	
	Fed state: storage of glucose glycogenesis	
	Fasting: production of glucose glycogenolysis, gluconcogenesis	
2. Protein	Catabolism	
	Krebs citric acid cycle	
	Krebs urea cycle	
	Synthesis	
	Albumin	
	Transferrin	
	Ceruloplasmin	
	Alpha-fetoprotein	
	Coagulation factors	
	Hormone-induced proteins	
3. Lipid	Synthesis	
	Cholesterol	
	Lipoproteins	
	Removal	
	Lipoproteins—LDL, VLDL	
4. Bilirubin	Conversion:	hemoglobin to bilirubin
		urobilinogen to bilirubin
	Conjugation:	to diglucuronide
	Secretion:	to plasma
5. Drug and toxin	Conversion:	lipid-soluble to water-soluble compounds
	Oxidation, hydroxylation: by phase 1 enzymes	
	Conjugation: by phase 2 enzymes	

patients with end-stage liver disease now provides an option for continued healthy and productive life.

2.1 Carbohydrates

The liver regulates the amount of fuel, or glucose, available to the body through its unique role in carbohydrate metabolism. Hepatocytes convert galactose to glucose; synthesize glycogen from glucose, lactic and pyruvic acids and glycerol; store glycogen; break down glycogen to glucose (glycogenolysis); and synthesize "new" glucose (gluconeogenesis) from amino acids and other precursors.

After ingestion of food, glucose is delivered to the liver via the portal vein and taken up into hepatocytes by facilitated diffusion. It is then irreversibly phosphorylated and trapped inside the liver cell by the enzyme glucokinase which is not saturated at physiologic levels of substrate. By acting as a "glucose sink," this process maintains tight homeostatic control of peripheral blood glucose levels (VAN THEIL et al. 1987).

Once phosphorylated, glucose can be: (a) stored as glycogen; (b) converted to fat and protein precursors; or (c) used as energy. Glycogen production in the hepatocyte is regulated by the concentration of glucose and its phosphorylated form through a phosphorylation-dephosphorylation enzymatic mechanism that in turn modulates the activity of enzymes involved in glycogen synthesis.

Hormones influence the metabolism of glucose and glycogen in the liver. Glucagon and epinephrine stimulate glycogen breakdown. Insulin increase the levels of glycogenic enzymes, but intrahepatic glucose levels are more important in regulating rapid changes in glycogen metabolism (BERGMAN 1977; DEWULF and HERS1967). Insulin does have a significant effect on overall hepatic glucose regulation. At low doses insulin inhibits glycogenolysis, while at higher levels insulin also inhibits gluconeogenesis in the hepatocyte (CHIASSON et al. 1980). The liver also acts as the major site of insulin degradation. After binding to specific receptors on the liver cell surface, insulin is internalized with subsequent breakdown by cytosolic proteases (GAMMELTOFT and TYGSTRUP 1983).

The liver then plays a major role in homeostasis of glucose which is the primary energy source for most tissues in the body. For example, after an overnight fast, peripheral blood insulin and portal venous glucose levels are low resulting in no hepatic uptake of glucose. Plasma glucose for tissue energy utilization is maintained by hepatic glucose production and release into the systemic circulation. The glucose is produced by the breakdown of glycogen and gluconeogenesis in the liver. Precursors for gluconeogenesis are derived predominantly from the breakdown of peripheral protein. As fasting is extended beyond 24 h, glycogen stores are depleted and most glucose is derived from hepatic gluconeogenesis. These pathways may be disrupted in severe liver disease resulting in major clinical consequences.

2.2 Proteins

The liver is the site of most protein degradation and synthesis for the human body. Some proteins are transaminated or deaminated to keto-acids which are then metabolized through other pathways such as the Krebs citric acid and urea cycles. In severe liver disease, 85% of the liver must be functionally impaired before blood and urinary amino acid levels increase. Urea production is also impaired, but hepatic reserves for this function are so great that low blood urea concentrations are only rarely observed in fulminant hepatic failure. A rise in blood ammonia level, which results from a failure of the Krebs citric acid cycle, is more common and is associated with the hepatic encephalopathy that often occurs in end-stage liver disease.

Many important plasma proteins are synthesized by the hepatocyte and secreted into the peripheral circulation. Dimunition in their levels during periods of liver dysfunction often has important clinical consequences. Included among the proteins synthesized by the liver are the acute phase response factors: albumin, fibrinogen, alpha-1-antitrypsin, heptaglobin, ceruloplasmin, and transferrin. They rise in response to tissue injury, which may contribute to maintaining serum concentrations of these proteins even in the face of liver disease.

Approximately 10 g albumin are synthesized by the liver each day. Albumin has a 22 d half life, and its serum concentration falls only gradually after onset of liver disease, even if quite severe. Alpha-1-antitrypsin is a genetically polymorphic trypsin inhibitor; the inheritance of a particular form of this protein results in a deficiency that is associated with liver and pulmonary disease. Haptoglobin is a glycoprotein and is lowered in hemolytic diseases and hepatocellular failure. Ceruloplasmin is the major copper-containing protein in the blood, and a low concentration is associated with Wilson's disease, a liver and brain disease that results from copper overload of those organs. Transferrin is the major iron transport protein. It is greater than 90% saturated in patients with hemochromatosis, a disease of iron overload in multiple organs, including the liver. Alpha-fetoprotein is a normal component of plasma in human fetuses and reaches a maximum concentration at about 16 weeks of fetal life. Shortly after birth it disappears from the blood. It is synthesized again by many hepatocellular carcinomas and is used clinically for the diagnosis and early detection of such tumors.

The liver is the principal site of synthesis of all coagulation factors except the von Willebrand factor and the fibrolytic factors (FAIR and MARLAV 1986; KELLY and SUMMERFIELD 1987). The liver also synthesizes protease inhibitors that modulate the coagulation cascade, such as antithrombin III, protein C, and heparin cofactor II. The liver uses vitamin K in the synthesis of many of the important blood coagulation factors. The half-life of these clotting proteins is short, and therefore they become depleted rapidly after the onset of hepatic dysfunction. Accordingly, other factors, such as thrombocytopenia, disseminated intravascular coagulation, and failure of bile salt secretion into the intestine resulting in inadequate absorption of vitamin K can all contribute to abnormal bleeding in patients with severe liver disease. Clinically, this is often manifested by bleeding from the gastrointestinal tract.

Different hormones exert a variety of different effects on the liver. Most hormones have a specific receptor on the cell surface or in the cytoplasm. Liver response to a given hormone depends on the concentration of the hormone as well as the number and characteristics of the receptors for the hormone. Membrane-bound hormones usually act through the adenylate cyclase system and convert adenosine triphosphate (ATP) to cyclic adenosine monophosphate (AMP) which is the intracellular mediator for hormone action. Cyclic AMP works through a protein kinase producing the characteristic response of the hepatocyte. Sex hormones, on the other hand, bind to receptors in the cell cytoplasm. The hormone receptor complex then is transported to the nucleus and influences gene expression (VAN THEIL et al. 1987).

2.3 Lipids

The liver plays a role in the metabolism of the major types of lipid carried in plasma, including cholesterol, phospholipids, triglycerides, and lipoproteins. Cholesterol, synthesized in the liver from acetate in the cytoplasm of hepatocytes, is the source of virtually all of the endogenously produced sterol. It is found in cell membranes and is a precursor of bile acids and steroid hormones. The genetically induced hypercholestolemias are the result of continued cholesterol synthesis by the liver in the face of high blood levels. Occasional patients with hepatocellular carcinoma will manifest the same defect and develop very high serum cholesterol levels. In liver disease associated with bile duct obstruction, total serum cholesterol rises and sometimes is deposited in the skin (xanthomas). In many hepatocellular diseases triglycerides tend to be elevated with the excess found in the low density lipoprotein (LDL) fraction (SHERLOCK 1989).

2.4 Bilirubin Metabolism

Bilirubin is the metabolic end product of hemoglobin. After breakdown of red blood cells in macrophages of the reticuloendothelial system, the heme portion of the hemoglobin molecule is catabolysed to biliverdin which is rapidly converted to bilirubin and transported to the liver. In the liver, bilirubin is conjugated to glucuronic acid or tightly bound to albumin and is then secreted into the bile. Bilirubin diglucuronide is the major conjugate in human bile. The secretion of bile, by hepatocytes, into the cannalicular system is controlled by at least two independent processes. One of these mechanisms is dependent on bile salts, while the other depends on organic anions such as bilirubin. Bilirubin glucuronide is polar and therefore not absorbed from the small intestine. In the colon it is broken down to urobilinogen which is nonpolar and readily absorbed. Reabsorbed urobilinogen is transported by the portal circulation to the liver, where it is taken up by hepatocytes and re-excreted into the biliary tree or into the plasma where it is carried in the general circulation to the kidney and excreted in urine. With advanced liver disease, re-excretion by the liver is impaired, and more urobilinogen ends up in the urine turning it dark green. Impairment of the re-excretion of urobilinogen is also responsible for symptoms of fever and heart failure in the early stages of viral hepatitis. Hepatocellular disease may result in jaundice by: disturbing the uptake and transport of bilirubin in the hepatocyte, causing a defect in conjugation, inducing abnormalities in the cell membrane opposite the canaliculus for excretion of bile, or by obstructing the large bile channels before they reach the intestine. Once an excess amount of bilirubin is in the circulation, clinical jaundice, including symptoms of dark skin and pruritis, results when bilirubin enters tissues. Circulating protein-bound bilirubin does not readily enter tissue fluids that are low in protein. Therefore, exudates tend to be more icteric than transudates. Skin, ocular sclera, and blood vessels have a high elastin content and easily become jaundiced (SHERLOCK 1989).

2.5 Drug and Toxin Metabolism

The liver is the principal organ involved in drug and toxin metabolism. Oral drugs that are lipid soluble pass the membrane of the intestine and are delivered to the liver via the portal system. In the liver they are converted to water-soluble compounds for excretion via urine or bile. Hepatic drug metabolism to such products can be divided into two phases. Phase 1 is the main drug-metabolizing system and resides in the smooth endoplasmic reticulum of the liver cell. The enzymes involved include the mixed function mono-oxygenases, cytochrome C reductase, and cytochrome P450. Reduced nicotinamide-adenine-dinucleotide phosphate (NADPH) is a cofactor. The drug is oxidated or hydoxylated to a more polar compound. Alternative phase 1 drug-metabolizing reactions include the conversion of alcohol to acetaldehyde by alcohol dehydrogenases. Phase 1 enzymes can be induced or inhibited by a variety of compounds.

Phase 2 metabolism of drugs involves polarization via conjugation with small endogenous molecules (e.g., glucuronide, sulfate). Although the enzymes catalyzing phase 2 reactions are found elsewhere in the body, they are usually most abundant in the liver. In patients with liver disease, the half-lives of drugs are often prolonged in correlation with the extent of hepatocellular damage (FARRELL et al. 1978). In these patients, side effects associated with excessive retention of certain hepatically cleared drugs may occur at much reduced dosages. Other drugs, which can cause liver toxicity by forming toxic intermediates, are also more likely to do so at a reduced dose in patients with already impaired liver function (SHERLOCK 1989).

3 Pathology

3.1 Patterns of Hepatic Injury

Hepatocytes are subject to injury by three distinct mechanisms: direct toxic effects of chemicals or metabolites; indirect, immunologically mediated damage; and the effects of bile retention (cholestasis) (SHERLOCK 1985). In humans with liver disease, all three mechanisms frequently operate simultaneously. This results, at the histopathologic level, in a mixture of lesions which makes assigning causes of liver injury by morphologic criteria alone virtually impossible.

Furthermore, hepatocytes have only a few ways of responding to injury. These are usually described in the light microscope as ballooning degeneration, eosinophilic (acidophilic) degeneration, hyaline degeneration, and fatty degeneration or steatosis. Ballooning is characterized by swelling of liver cells, clear or finely granular cytoplasm, and enlarged, irregular, pale-staining nuclei. In the electron microscope, the cisternae of the endoplasmic reticulum are dilated, the cytoplasm is degranulated, and the glycogen content is reduced. Balloon cells are thought to have a very limited survival, rapidly undergoing lytic necrosis, breakdown of cell membranes, and disappearance. Ballooning is commonly seen in association with immune-mediated damage and cholestasis.

In eosinophilic degeneration, the affected hepatocytes are shrunken rather than swollen; the cytoplasm is dark pink (eosinophilic) with hematoxylin and eosin (H and E) stains; and the nuclei are shrunken, darkly stained, and structureless (pyknotic). These cells continue to shrink, lose their nuclei, and are extruded from liver cell plates into sinusoidal spaces. There they may persist as deeply eosinophilic-staining bodies of varying sizes and shapes. These structures were first described by Councilman (and are often called "Councilman bodies") (ISHAK 1976).

Hyaline degeneration and steatosis are best understood in the context of direct injury to the liver. In this process, the damage is primarily to cytoplasmic organelles including mitochondria, endoplasmic reticulum, and microtubules. Alcohol is the classic example of a directly toxic agent. Metabolites of alcohol (e.g., acetaldehyde) damage the mitochondrial membrane. In the electron microscope the mitochondria are enlarged and distorted. Hyaline degeneration is frequently indicated in pathologist's reports by the presence of Mallory bodies (also called alcoholic hyaline) (MALLORY 1911). These bodies appear in the light microscope as eosinophilic masses of hyaline which may occupy most of the cytoplasm. Ultrastructurally, Mallory bodies are composed of filaments of varying sizes and shapes. These, in turn, are probably related to abnormalities in the synthesis and function of microtubules (SHERLOCK 1985).

Steatosis, the accumulation of fat in hepatocytes, is often prominent in alcoholic liver disease. There are several metabolic abnormalities that contribute to the accumulation of fat including increased synthesis of triglycerides and lipoproteins, lipid peroxidation, and decreased activity of the citric acid cycle (NILIUS 1985).

These changes in hepatocytes are accompanied by a cellular infiltrate of polymorphonuclear leukocytes and proliferation of Kupffer cells. Taken together, these abnormalities are representative of the spectrum of effects of direct liver injury. None of these abnormalities is specific for alcoholic liver disease. Any or all of them may occur in other, widely disparate diseases such as diabetes mellitus, obesity, and Wilson's disease (a disorder of copper metabolism) (SHERLOCK 1985).

Viral hepatitis demonstrates the indirect, immune mechanism of damage to hepatocytes. The injury is to the cell membrane and is due to specific and nonspecific cytotoxic effects of lymphocytes. With the light microscope, the most prominent abnormality is a rich infiltrate of lymphocytes and plasma cells in portal areas and within lobules. Liver cells show evidence of balloon degeneration, eosinophilic degeneration, and necrosis as well as regeneration (BIANCHI et al. 1979). The stimulus to regeneration is cell death. Thus, hepatocyte necrosis is always accompanied by increased numbers of cells in mitosis, the hallmark of regeneration. In the normal mature liver, only one cell in 10000 is in mitosis (BECKER 1974).

Cholestatic injury is characterized by arrested bile and, at the light microscopic level, minimal hepatocyte damage, mainly accumulation of bile in liver cells and in canaliculi. Ultrastructurally, however, the lesions of cholestasis are not trivial. The Golgi apparatus is vacuolated, lysosomes are increased in number, and the endoplasmic reticulum is hypertrophied. One hypothesis is that these changes are brought about by the retention of bile salts. Cholestasis can occur as the result of either mechanical obstruction to the outflow of bile (e.g., a stone in the common bile duct) or as a functional impairment of the secretion of bile by hepatocytes. Cholestasis may occur as a response to drugs or chemical exposures, or as part of the spectrum of virus-induced lesions (SHERLOCK 1985).

4 Pathology of Liver Diseases Associated with Hepatitis B

4.1 Acute Hepatitis

The term "acute hepatitis" refers to a disease characterized by diffuse inflammation of the liver. It is not synonymous with "transient" hepatitis. Hepatitis viruses A,B,C, and E can all cause acute hepatitis. The disease is always transient when caused by the A or E viruses, but can be either transient or persistent when induced by the B or C viruses. Even though only acute hepatitis B or C may progress to chronic forms of hepatitis, the histopathology of acute viral hepatitis is indistinguishable whether caused by hepatitis viruses A, B, C, or E. The entire liver is involved, usually to the same degree in all areas. Four morphological patterns may occur that refer to the location and extent of hepatocellular necrosis and are, hence, approximately related to the severity of the disease. They are: classical hepatitis with spotty necrosis, hepatitis with bridging necrosis, with panlobular necrosis, or with periportal necrosis (SCHEUER 1985).

In classical acute hepatitis, hepatic damage is focal. Liver cell necrosis involves single hepatocytes or small groups of cells, particularly in the centers of lobules. Balloon degeneration, eosinophilic degeneration, and Councilman bodies can be detected. Portal tracts and sinusoids are filled initially with lymphocytes and later, as the lesion evolves, with a mixed population of inflammatory cells including lymphocytes, polymorphonuclear granulocytes, and monocytes. The phagocytic cells scavenge the necrotic debris, but they may also contribute to the damage to hepatocytes by releasing the products of oxidative metabolism (BIANCHI et al. 1979; SCHEUER 1985).

Degenerative and regenerative changes are both present in lobules. Liver cells vary in size and shape (pleomorphism) with several mitoses and binucleated cells seen in each high-powered field. There is diffuse enlargement of sinusoidal lining cells and focal areas of hyperplasia. It is thought that the lining cells are released into the sinusoids and become macrophages.

Bile pigment is not always present. When it is, it can occur as bile thrombi in canaliculi, as fine or coarse granules in hepatocytes, or within Kupffer cells.

As the acute stage of hepatitis resolves, the liver shows a restoration of normal parenchymal architecture. Some areas of spotty necrosis may persist, along with inflammatory cells in the portal areas and around hepatic veins in lobules. Fine reticular fibers between portal and central zones and between portal tracts may be seen at this stage, but they do not progress to cirrhosis (BIANCHI et al. 1979).

In acute hepatitis with bridging necrosis, the extent of hepatocyte damage is greater than in classical acute hepatitis. There are bands of confluent necrosis between centrilobular areas and portal tracts (BOYER and KLATSKIN 1970). According to RAPPAPORT'S (1976) description of the microanatomy of the hepatic acinus, zone 3 receives the poorest blood supply from the terminal branches of the portal vein and the hepatic artery. It is this zone of relatively poor blood supply that is most affected by bridging necrosis. Balloon degeneration with rapid lysis is the major hepatocyte lesion, which results in the "dropping out" of hepatocytes from hepatic plates. The bands of necrosis may vary in width, cellularity, and presence or absence of septae. The cellular infiltrate, which is predominantly composed of phagocytic cells (polymorphonunclear leukocytes and monocytes) rather than lymphocytes, may reflect the speed of necrosis rather than the mechanism of injury.

Clinically, the significance of bridging necrosis is that it carries a much greater risk of a poor outcome than classical hepatitis with spotty necrosis. Although in some cases it may heal and leave a normal liver, more frequently, in the context of HBV or HCV infection, it may result in chronic hepatitis, cirrhosis, or death from hepatic failure (BOYER and KLATSKIN 1970).

Acute hepatitis with panlobular (massive) necrosis is an even more severe lesion. The necrosis involves entire acini or entire lobules. When the whole liver is affected, it is recognized clinically as fulminant hepatitis and carries a 70% risk of death. Survival apparently depends on rapid liver cell regeneration and is reflected clinically by a rise in the serum alpha-fetoprotein concentration (KARVOUNTZIS and REDEKER 1974). Fortunately, panlobular necrosis is relatively rare; less than 1% of cases of acute hepatitis become fulminant.

Periportal necrosis is characterized by extensive cellular infiltration and necrosis in periportal and portal areas out of proportion to other areas. There may be portal to portal bands of necrosis. POPPER et al. (1965) described necrosis of liver cells at the interface of parenchymal and connective tissue and gave it the name "piecemeal necrosis." This type of necrosis is associated with, and may give rise to, periportal necrosis. Patients with periportal necrosis infected with HBV or HCV are at high risk of developing chronic hepatitis and cirrhosis (BIANCHI et al. 1979).

Hepatitis B viral antigens, when present, can be detected in either frozen sections or formalin-fixed, paraffin-embedded liver tissues by immunofluorescence or immunoperoxidase methods. In classical acute hepatitis B, it is frequently difficult to detect viral markers in tissues, particularly at the height of the disease. An occasional cell may stain positively for core antigen (HBcAg) in its nucleus. It is thought that the absence of viral antigens reflects the active elimination of infected cells by the host's immune response. In the prenecrotic stage of acute hepatitis, surface antigen (HBsAg) may be detected in cytoplasm and plasma membranes and HBcAg in nuclei of hepatocytes (GERBER and THUMS 1985). Elution of isolated plasma membranes of hepatocytes from patients with acute hepatitis B has demonstrated antibodies to core, but not to surface antigen, implying that HBcAg is the more likely target of the immune response (MONDELLI et al. 1982). Also, the binding of anti-core antibodies to the cell surface may prevent the demonstration of HBcAg on the plasma membrane by immunohistochemistry.

In the more severe types of acute hepatitis, HBcAg may be present in the nuclei of up to 60% and HBsAg in the cytoplasm or cell membrane of 30% of hepatocytes (GUDAT and BIANCHI 1977). Usually the same cells do not stain positively for both core and surface antigens, although this observation may be more a problem of detection than of presence (GOWANS and BURRELL 1985). This distribution of HBV antigens has been called the "focal HBcAg type" (GUDAT 1975). The ability to detect viral antigens in more aggressive cases of acute hepatitis, but not in classical acute hepatitis, is compatible with the hypothesis that failure of the immune response to eliminate many of the HBV-infected cells is responsible for both the severity of the illness and the high propensity to progress to chronic hepatitis (BIANCHI et al. 1979).

4.2 Chronic Hepatitis

Chronic hepatitis is defined as inflammation of the liver that has persisted without improvement for 6 months (SCHEUER 1979). With respect to both HBV and HCV, this definition may be difficult to apply because the duration of liver disease may not be known when the patient is first seen. The development of chronic hepatitis with these viruses usually occurs insidiously following a mild or clinically inapparent acute hepatitis.

Chronic hepatitis involves both inflammation of the liver and damage to hepatocytes. The various ways of categorizing chronic hepatitis are aimed at distinguishing cases with significant parenchymal damage (chronic active hepatitis, chronic aggressive hepatitis) from those with inflammatory infiltrates but minimal cell damage (chronic persistent hepatitis, chronic lobular hepatitis). As with the types of acute hepatitis, these are histopathologic diagnoses made by the examination of liver biopsies. Chronic hepatitis can be caused by hepatitis viruses B and C, but apparently not A or E. Again, the lesions of chronic hepatitis can also be induced by nonviral agents, e.g., alcohol, drugs, copper (Wilson's disease), iron (hemochromatosis), autoimmunity.

Chronic active hepatitis (chronic aggressive hepatitis, CAH) may occur with or without bridging necrosis and with or without cirrhosis. When present without bridging necrosis, the inflammatory infiltrates, which are rich in lymphocytes, are confined to periportal areas and extend into the adjacent parenchyma. The lobular architecture is preserved, but hepatocytes are destroyed by the process of piecemeal necrosis. Regeneration accompanies the continuing cell destruction with the formation of foci of new cells that are ultimately recognized as distinct nodules. Cirrhosis may develop if collagen fibers are laid down in the portal tracts, and nodular regeneration becomes prominent (SCHEUER 1979).

CAH with bridging necrosis is a more severe lesion that tends to progress to cirrhosis more rapidly than in the absence of bridging necrosis. Both piecemeal necrosis and lymphocytic infiltration are present with the bridging necrosis. The lobular architecture is distorted by bands of confluent necrosis connecting portal to centrilobular areas and extensive infiltration of lymphocytes within lobules. Regenerating cells often form gland-like clusters (rosettes). Balloon degeneration and hepatocyte swelling are commonly seen. The transition to cirrhosis is not sharply demarcated, and chronic active hepatitis with cirrhosis is frequently encountered.

The distribution of HBV antigens in CAH is of the focal HBcAg type (see Sect. 4.1). Both nuclear HBcAg and cytoplasmic and membrane-associated HBsAg are readily detected, with about twice as many liver cells containing core as surface antigens. In general, the greater the inflammatory response, the fewer the core positive cells detected. As in the severe forms of acute hepatitis, the inability of the immune response to remove all virus-infected cells results in continuing cell death and continuing infection (GUDAT 1975). A further support for this notion is the observation that in patients with CAH who improve with alpha-interferon therapy, a period of increased cell necrosis is followed by disappearance of viral antigens from the liver. Patients who do not improve with interferon treatment demonstrate persistence of viral antigens in liver biopsies (BROOK et al. 1988).

Chronic persistent hepatitis (CPH) is characterized by cellular infiltration and enlargement of portal tracts with minimal piecemeal necrosis of parenchymal cells. This lesion is not specific for viral hepatitis. Many systemic inflammatory diseases will demonstrate portal infiltrates on liver biopsy. Therefore, the diagnosis of viral CPH requires serologic or other specific evidence of chronic infection with hepatitis B or C viruses.

The lobular architecture is preserved, and the parenchymal cells show little or no damage. Cirrhosis does not develop unless there is first progression to CAH.

In the usual patient with CPH, few cells with HBcAg are detected and many HBsAg positive cells are seen. An exception to this pattern are patients who are immunosuppressed either by virtue of an underlying disease (e.g., end-stage renal disease, AIDS, etc.) or from treatment with immunosuppressive drugs (corticosteroids, cyclosporin). Liver biopsies from such patients reveal HBcAg in virtually every hepatocyte nucleus (GUDAT 1975). HBsAg is generally detected in a lower percentage of cells and is often distributed diffusely on liver cell membranes.

Chronic lobular hepatitis (CLH) is histopathologically an even milder lesion than CPH. It is similar to classical acute hepatitis with spotty necrosis in that lobules are infiltrated with lymphocytes, necrosis is focal and mild, and portal infiltration is not prominent. Progression to CAH or cirrhosis is unusual. The distribution of viral antigens is similar to that seen in CPH.

4.3 Cirrhosis

Cirrhosis is a type of host response to chronic liver injury that is characterized by bands of fibrosis throughout the liver associated with and surrounding nodules of parenchymal tissue. The initiating event is necrosis of hepatocytes. In "active" cirrhosis there is continuing liver cell damage and death which provides the stimulus for hepatocyte regeneration, nodule formation, and laying down of more fibrous tissue. "Inactive" cirrhosis indicates that hepatocyte damage has ceased or has been reduced to a low level, resulting in arrest of the other components of the disease process (SCHEUER 1979). As with acute and chronic hepatitis, cirrhosis is not a specific manifestation of chronic viral infection. It can be caused by drugs, alcohol, iron (hemochromatosis), retention of bile constituents (biliary cirrhosis), as well as chronic hepatitis B or C.

Piecemeal necrosis is commonly seen in active cirrhosis of whatever cause, as well as cholestasis, steatosis, and iron accumulation. Nodule formation is the result of over-activity of the regeneration process. That is, the stimulus to cell division is cell death. If the replacement process were precise, there would be a one-to-one exchange of dead cells for new cells. In cirrhosis, however, cell division usually continues beyond replacement giving rise to small (less than 3 cm in diameter) or large hyperplastic nodules (micro- or macronodular cirrhosis), which lack the normal association into hepatic plates (ANTHONY et al. 1978). HBV and HCV are usually associated with macronodular and alcohol abuse with micronodular cirrhosis.

Cirrhosis in its early stages is usually asymptomatic, but with time patients develop effects that are often fatal. As nodules form and grow, they distort the architecture of lobules causing compression of hepatic venules. Fibrosis in portal areas may obstruct portal venous blood flow. Both fibrosis and nodule formation increase resistance to blood flow through sinusoids. Shunts may develop between portal venules and hepatic arterioles. These changes increase the pressure in the portal vein and its tributaries (portal hypertension) which in turn causes the spleen to enlarge (splenomegaly) and to induce dilatation of small veins at the lower end of the esophagus and upper end of the stomach (esophageal varices) (SCHEUER 1979). These veins are thin walled and are prone to rupture causing massive hemorrhage. Rupture of esophageal varices is a frequent cause of death of patients with cirrhosis.

A second cause of morbidity and mortality in cirrhosis is dysfunction of hepatocytes. The various vascular changes in the liver reduce access of liver cells to nutrients and oxygen. Continuing hepatocyte necrosis eventually reduces the number of functioning hepatocytes. These abnormalities result in a decreased synthesis of albumin which, in turn, decreases the intravascular osmotic pressure. Portal hypertension and low osmotic pressure result in fluid losses from mesenteric capillaries and leakage of hepatic lymph into the peritoneal cavity (ascites). Failure of hepatocytes to conjugate and excrete bile leads to an accumulation of bilirubin in the circulation and in the skin (jaundice). The failure of the liver to clear various metabolites, particularly ammonia, from the circulation may result in disorientation, coma, and death (MACSWEEN and SCOTT 1973).

HBV antigens have an interesting distribution in cirrhotic livers. Each nodule appears to have its own pattern. Some nodules are free of viral markers, while in others the cytoplasm of every cell stains intensely and homogeneously for surface antigen. Some nodules show a uniformly membranous distribution of HBsAg with little cytoplasmic staining. HBcAg is less commonly demonstrated and, when seen, is usually in nuclei (NAYAK et al. 1977; NAZAREWICZ et al. 1977). In some nodules however, all cells may have a cytoplasmic distribution of core antigen (ZHOU et al. 1987). This heterogeneity of expression may reflect different degrees of maturation of hepatocytes among regenerating nodules.

5 Hepatocellular Carcinoma

Patients with cirrhosis of any etiology, but particularly those chronically infected with HBV or HCV, are at increased risk of developing a cancer in the liver. These tumors are often called "primary" hepatocellular carcinomas (PHC) to clearly distinguish them from tumors that arise in other organs and metastasize to the liver.

According to the World Health Organization, about 80% of the cases of PHC in the world are associated with chronic HBV infection (WHO 1983). Also, about 80% of PHCs arise in a setting of cirrhosis. In another 15% of cases the liver is affected by chronic hepatitis (CAH or CPH). Less than 5% of cases occur in a histologically normal liver (LONDON 1981).

In the past it was thought that most PHCs were multicentric (i.e., multiple primary tumors). Recent studies of patients with PHCs, diagnosed at an early stage, indicate that most or all of these tumors begin as solitary nodules (TANG 1985). The tumors consist of parenchymal cells which may be well, moderately, or poorly differentiated. The tumor cells are often arranged in trabeculae two to eight cells thick, separated by sinusoids. Kupffer cells are usually absent from tumor nodules (PETERS 1976).

PHCs are highly lethal, usually causing the death of the patient within 6 months of diagnosis. The tumors commonly invade the portal and venous systems allowing rapid spread throughout the liver. Death is generally the result of hepatic failure rather than from metastases to other vital organs.

Viral antigens are distributed differently in the cancerous and noncancerous parts of livers containing PHC. Because PHCs are usually associated with cirrhosis, the distribution of HBsAg and HBcAg in the nontumor areas is the same as that described above in cirrhotic liver. Both surface and core antigens are less frequently observed in tumor than nontumor tissue (NAYAK et al. 1977; CAO 1985). When HBsAg is detected, it is usually sparsely distributed with more intense staining perinuclearly. Surface antigen is more frequently seen in well-differentiated than in poorly differentiated tumors. HBcAg found in tumor cells in less than 10% of cases. In the few instances where it is detected, it may have a cytoplasmic distribution (ZHOU et al. 1987).

6 Healthy Carriers of HBV

The term "chronic carrier" and "healthy carrier" are commonly used to describe individuals chronically infected with HBV, but without symptoms or signs of chronic liver disease. In practice these words are used loosely and may sometimes include all chronically infected persons whether they have liver disease or not. Here we wish to focus on chronically infected individuals without liver disease. Such persons are identified in serologic surveys of general populations or by the screening of blood donors by blood banks. As many as half of all chronically infected persons may be healthy carriers.

Liver biopsies have been performed on many healthy carriers in various parts of the world. Most commonly, minimal changes of chronic lobular hepatitis or, less often, chronic persistent hepatitis have been reported (GERSTLEY et al. 1972; SCHEUER 1979). The most common histopathologic characteristic of the livers of chronic carriers is the presence of large numbers of "ground-glass" cells (HADZIYANNIS et al. 1973). These are large hepatocytes with homogeneous, pale pink cytoplasm. Specific immunohistologic stains show that ground-glass cells stain strongly for HBsAg, but negatively for HBcAg (GERBER et al. 1975). At the ultrastructural level, these cells reveal proliferation of the endoplasmic reticulum with many 20-nm spherical and tubular structures in their cisternae (TANIKAWA 1979).

Most clinical investigators believe that healthy carriers are at much lower risk of developing liver cancer than chronically infected persons with chronic active hepatitis or cirrhosis, thus emphasizing the importance of liver damage in hepato-carcinogenesis.

References

Anthony PP, Ishak KG, Nayak NC, Poulsen HC, Scheuer PJ, Sobin LH (1978) The morphology of cirrhosis. J Clin Pathol 31: 395–414
Arey LB (1932) On the presence of so called portal lobules in the seal's liver. Anat Rec 51: 315–322
Becker FF (1974) Regeneration. In: The biochemistry of disease, vol 5, The liver, normal and abnormal functions. Dekker, New York

Bergman RN (1977) Integrated control of hepatic glucose metabolism. Fed Proc 36: 265–270
Bianchi L, Zimmerli-Ning M, Gudat F (1979) Viral hepatitis. In: Pathology of the liver. Churchill Livingstone, Edinburgh
Blum HE, Stowring L, Figus A, Montgomery CK, Haase AT, Vyas GN (1983) Detection of hepatitis B DNA in hepatocytes, bile duct epithelium, and vascular elements by in situ hybridization. Proc Natl Acad Sci USA 80: 6685–6688
Boyer JL, Klatskin G (1970) Pattern of necrosis in acute viral hepatitis. Prognostic value of bridging (subacute hepatic necrosis). N Engl J Med 283: 1063–1071
Brittain RS, Marchioro TL, Hermann G, Waddell WR, Starzl TE (1964) Accidental hepatic artery ligation in humans. Am J Surg 107: 822–832
Brook MG, Petrovic L, MacDonald JA, Scheuer PJ, Thomas HC (1988) Evidence of histological improvement following antiviral treatment in chronic hepatitis B virus carriers and identification of histological features that predict response. In: Zuckerman AJ (ed) Viral hepatitis and liver disease. Liss, New York
Cao YZ (1985) Hepatitis B virus and hepatocellular carcinoma: viral antigens HBsAg and HBcAg in primary hepatocellular carcinoma tissue. In: Tang ZY (ed) Subclinical hepatocellular carcinoma Springer, Berlin Heidelberg New York
Chiasson JL, Atkinson RL, Cherrington AD, Cherrington U, Kellar U, Sinclair-Smith BC, Lacy WW, Lijenquist JE (1980) Effects of insulin at two-dose levels on gluconeogenesis from alanine in fasting men. Metabolism 29: 810–818
DeWulf H, Hers HG (1967) The stimulation of glycogen synthesis and of glycogen synthetase in the liver by administration of glucose. Eur J Biochem 2: 50–56
Elias H (1949) The liver cord concept after one hundred years. Science 110: 470–472
Erlinger S (1987) Secretion of bile. In: Schiff L, Schiffe (eds) Diseases of the liver. Lippincott, Philadelphia
Fair DS, Marlar RA (1986) Biosynthesis and secretion of factor VII, protein C, Protein S, and the protein C inhibitor from a human hepatoma cell line. Blood 67: 64–70
Farrell GC, Cooksley WGE, Hart P (1978) Drug metabolism in liver disease: identification of patients with impaired hepatic drug metabolism. Gastroenterology 75: 580–588
Gammeltoft S, Tygstrup N (1983) Carbohydrate metabolism in relation to liver physiology and disease. In: Arias IM, Franke M, Wilson JMP (eds) The liver annual 3. Elsevier, Amsterdam
Gerber MA, Thung SN (1985) Molecular and cellular pathology of hepatitis B. Lab Invest 52: 572–590
Gerber MA, Hadziyannis S, Vernace S, Vissoulis C (1975) Incidence and nature of cytoplasmic hepatitis B antigen in hepatocytes. Lab Invest 32: 251–256
Gerstley BJS, Custer RP, Blumberg BS, London WT, Sutnick AI, Coyne VZ (1972) Liver biopsies in patients with and without Australia antigen. Arch Pathol 93: 366–371
Gowans EK, Burrell CJ (1985) Widespread presence of cytoplasmic HBcAg in hepatitis B infected liver detected by improved immunochemical methods. J Clin Pathol 38: 393–398
Grisham JW (1976) Scanning electronmicroscopy of normal rat liver the surface structure of its cells and tissue components. Am J Anat 144: 295–301
Gudat F, Bianchi L (1977) Evidence for phasic sequences in nuclear HBcAg formation and cell membrane-directed flow of core particles in chronic hepatitis B. Gastroenterology 73: 1194–1197
Gudat F, Bianchi L, Sonnabend W, Thiel C, Aemshaenslin W, Stalder GA (1975) Pattern of core and surface expression in liver tissue reflects state of specific immune response in hepatitis B. Lab Invest 32: 1–9
Hadziyannis S, Gerber MA, Vissoulis C, Popper H (1973) Cytoplasmic hepatitis B antigen in 'ground glass' hepatocytes of carriers. Arch Pathol 96: 327–330
Hoffman AF (1989) The enterohepatic circulation of bile acids in health and disease. In: Sleisenger MH, Fondtran JS (eds) Gastrointestinal disease. Saunders, Philadelphia
Ishak KG (1976) Light microscopic morphology of viral hepatitis. Am J Clin Pathol 65: 787–827
Karvountzis GG, Redeker AG (1974) Relation of alpha-fetoprotein in acute hepatitis to severity and prognosis. Ann Intern Med 80: 156–160
Kelly DA, Summerfield JA (1987) Hemostasis in liver disease. Semin Liver Dis 7: 182–191
Kiernan F (1833) The anatomy and physiology of the liver. Philos Trans R Soc London 123: 711–710
London WT (1981) Primary hepatocellular carcinoma—etiology, pathogenesis, and prevention. Hum Pathol 12: 1085–1097
MacSween RNM, Scott AR (1973) Hepatic cirrhosis: a clinico-pathological review of 520 cases. J Clin Pathol 26: 936–942
Mall FP (1906) A study of the structural unit of the liver. Am J Anat 5: 227–308

Mallory FB (1911) Cirrhosis of the liver, five different lesions from which it may arise. Bull Johns Hopkins Hosp 22: 69–75
Mondelli M, Vergani GM, Alberti A, Vergani D, Portmann B, Eddelston ALWF (1982) Specificity of T-lymphocyte cytotoxicity to autologous hepatocytes in chronic hepatitis B infection. Evidence that T cells are directed against HBV core antigen expressed on hepatocytes. J Immunol 129: 2773–2778
Nayak NC, Dhan A, Sachdeva R, Mittal A, Seth HN, Sudarsanam D, Reddy B, Waghaolikar UL, Reddy CRRM (1977) Association of human hepatocellular carcinoma and cirrhosis with hepatitis virus surface and core antigens in the liver. Int J Cancer 20: 643–654
Nazarewicz T, Krowczynski K, Slusarczyk J, Nowoslawski A (1977) Cellular localization of hepatitis B virus antigen in patients with hepatocellular carcinoma coexisting with liver cirrhosis. J Infect Dis 153: 298–302
Nilius R (1985) Acetaldehyde, aldehyde dehydrogenase, and pathogenetic aspects of alcoholic liver disease. In: Brumer H, Thaler H, Festschrift for Hans Popper. Raven, New York
Opie EL (1944) The pathogenesis of liver tumors produced by butter yellow. J Exp Med 80: 231–246
Peters RL (1976) Pathology of hepatocellular carcinoma In: Okuda K, Peters RL (eds) Hepatocellular carcinoma, ed. Wiley, New York
Popper H, Paranetto F, Schaffner F (1965) Immune processes in the pathogenesis of liver disease. Ann NY Acad Sci 124: 781–799
Rappaport AM (1953) In discussion of Elias H: morphology of the liver, Trans 11th Liver Injury Conf, Macy, New York
Rappaport AM (1976) The microcirculatory acinar concept of normal and pathological hepatic structure. Beitr Pathol 157: 215
Rappaport AM (1987) Physioanatomic considerations. In: Schiff L, Schiff E (eds) Diseases of the liver. Lippincott, Philadelphia
Rappaport AM, Borowy ZJ, Lougheed WM, Lotto WN (1954) Subdivision of hexagonal liver lobules into a structural and functional unit: role in hepatic physiology and pathology. Anat Rec 119: 11–27
Scheuer PJ (1979) Chronic hepatitis. In: MacSween RNM, Anthony PP, Scheuer PJ (eds) Pathology of the liver: Churchill Livingstone, Edinburgh London New York
Scheuer PJ (1985) Pathological aspects of viral hepatitis. In: Brummer H, Thaler H (eds) Hepatology, Festschrift for Hans Popper. Raven, New York
Sherlock S (1985) Patterns of hepatic injury in man. In: Brunner H, Thaler H (eds) Hepatology, Festschrift for Hans Popper. Raven, New York
Sherlock S (1989) Anatomy and function. In: Sherlock S (ed) Diseases of the liver and biliary system. Blackwell Scientific, Cambridge
Tanikawa K (1979) Virus particles. In: Ultrastructural aspects of the liver and its disorders, 2nd edn. Igakushoin, Tokyo
Tang ZY (1985) Subclinical hepatocellular carcinoma-historical aspects and general considerations. In: Tang ZY (ed) Subclinical hepatocellular carcinoma. Chinese Academic, Beijing, Springer-Verlag, Berlin
Van Theil D, Stone BG, Schade RR (1987) The liver and its effect on endocrine function in health and diseases. In: Schiff L, Schiff E (eds) Diseases of the liver. Lippincott, Philadelphia
WHO Scientific Group on Prevention and Control of Hepatocellular Carcinoma (1983) Prevention of primary liver cancer. Report of a meeting of a WHO scientific group. Lancet 1: 463–465
Zhou X-D, DeTolla L, Custer RP, London WT (1987) Iron, ferritin, hepatitis B surface and core antigens in the livers of Chinese patients with hepatocellular carcinoma. Cancer 59: 1430–1437

Transcriptional Control of Hepadnavirus Gene Expression

H. Schaller and M. Fischer

1 Introduction

Hepadnaviruses multiply their DNA genome by reverse transcription of an RNA intermediate, termed pregenomic RNA. Therefore, hepadnaviral transcription has a dual function: the production of the RNA template for genome replication and the synthesis of messenger RNAs (mRNAs). The very small viral genome is fully covered by extensively overlapping coding sequences, requiring an expression strategy that uses differential transcription initiation and unspliced transcripts to produce at least seven viral gene products from four open reading frames. This complex organization implies that the hepatitis B virus (HBV) genome contains several transcriptional control regions, that these regions are located at short distances from each other (maximally 1.6 kb apart in the circular 3.2-kb genome), and that all are themselves actively transcribed during viral replication and eventually translated into protein. This scenario,

---

Zentrum für Molekulare Biologie, University of Heidelberg, Im Neuenheimer Feld 282, 6900 Heidelberg, F.R.G.

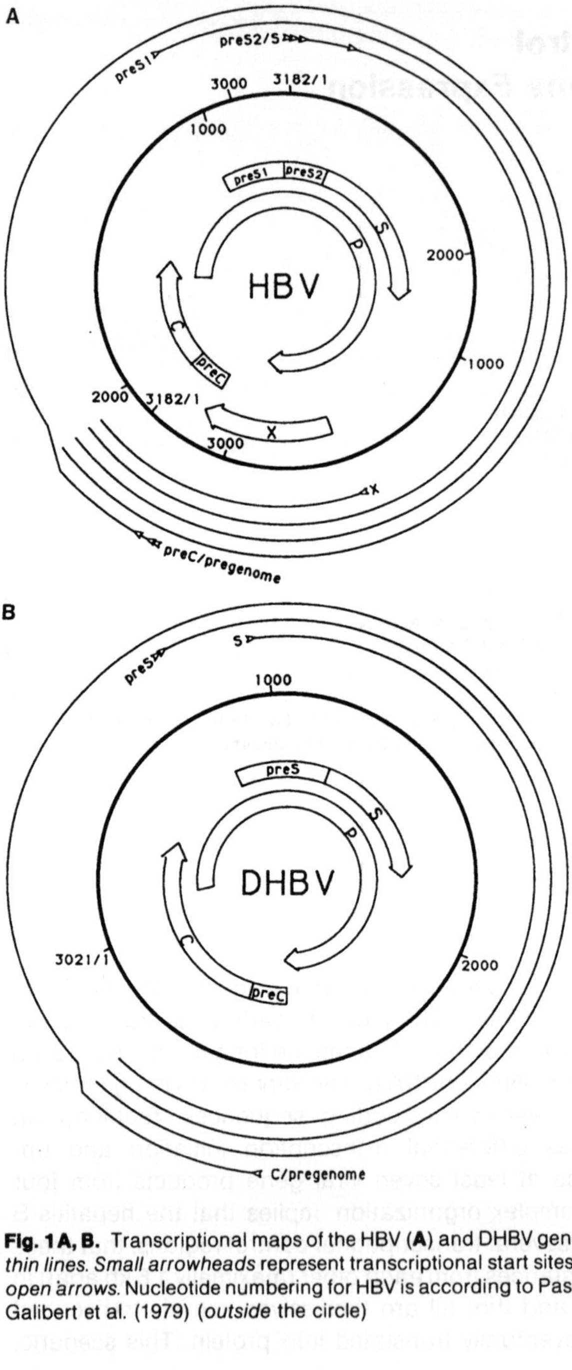

Fig. 1A, B. Transcriptional maps of the HBV (**A**) and DHBV genomes (**B**) The RNAs are displayed as *thin lines. Small arrowheads* represent transcriptional start sites. Open reading frames are drawn as *open arrows.* Nucleotide numbering for HBV is according to Pasek et al. (1979) (*inside* the circle) and Galibert et al. (1979) (*outside* the circle)

which is without precedent in other known replication units, has most likely imposed unusual restrictions on the evolution of the mechanisms regulating hepadnaviral gene expression and hence on elements involved in transcriptional control.

In addition to these difficulties arising from the necessity of expressing an optimal repertoire of gene products from a small, highly compact genome, there are also major technical limitations which have restricted our knowledge of hepadnaviral transcription for a long time. First, most studies were, and still are, being performed with HBV as the prototypic hepadnavirus; however, for this virus, to date no cell culture system exists that is permissive for all steps of the replication cycle. Secondly, hepadnaviral transcription is highly tissue specific and, therefore, early results derived from transfection of cultured cells could only be believed, with some confidence, to be related to the in vivo situation after certain differentiated hepatoma cell lines had been identified which supported production of infectious virus from cloned viral DNA genomes (see chapter by C. SEEGER et al. this volume). Consequently, much of our knowledge of HBV transcription was initially built on the physical mapping of the RNA species produced by animal hepadnaviruses in the liver of their respective hosts (BÜSCHER et al. 1985; ENDERS et al. 1985; MÖRÖY et al. 1985) and on analogous results from a single chimpanzee infected by an HBV isolate of known nucleotide sequence (CATTANEO et al. 1983, 1984; WILL et al. 1987). Much additional information on the transcriptional potential of the hepadnaviruses has been gained from transfection experiments with constructs containing all, or parts of cloned viral DNA genomes. This approach, in conjunction with site directed mutagenesis, has identified numerous *cis*-acting control elements as well as the exact position of many transcription initiation sites in the viral genome, and has also helped to clarify the structure and coding capacity of individual transcripts.

As a result of these studies a basic pattern of hepadnaviral transcription has emerged, which is now generally accepted as a central part of the viral life cycle. No significant differences are known between HBV and the other two mammalian viruses studied in detail, the woodchuck hepatitis virus (WHV) and the ground squirrel hepatitis virus (GSHV), and, as outlined in Fig. 1B, a very similar general transcriptional strategy was found with duck hepatitis B virus (DHBV), the avian prototypic virus. In this article we will therefore focus primarily on the better-studied HBV system. Data from the mammalian animal virus models will be cited only if they provide additional information, and for the avian viruses only relevant differences from the mammalian viruses will be pointed out.

2 The HBV Transcripts

Extensive circumstantial evidence indicates that viral transcripts are produced from a low number of covalently closed circular viral DNA genomes utilizing the nuclear transcription machinery of the host cell to produce a set of unidirectional and overlapping transcripts. As outlined in Fig. 1A and Table 1, four classes of mRNAs have so far been identified for HBV (three in the avian virus DHBV; Fig. 1B, Table 2) and correlated with specific functions in the viral life cycle. These transcripts initiate staggeredly on the circular DNA genome from a corresponding set of promoters and terminate at a single, common polyadenylation site. All RNAs are unspliced and, therefore, the shorter RNAs are fully contained in the larger species, the latter being necessarily polycystronic. However, as with cellular mRNAs, only the first open reading frame (ORF) is efficiently translated in each transcript, and these transcripts are often named accordingly.

Two highly abundant classes of transcripts are produced in the acutely infected liver by all hepadnaviruses. These are the terminally redundant genomic transcripts with a size of about 3.5 kb, and the subgenomic transcripts of about 2.1 kb. Consistent with their abundance, both serve to provide the major

Table 1. HBV transcripts

Apparent size[a] (kb)	Start sites[b]	Calculated length[c] (without polyA) (kb)	Function
3.5	1783/1784 1790 ± 1	3.35	preC mRNAs
3.5	1818	3.30	C mRNA pregenome
2.4	2807	2.30	preS1 mRNA
2.1	3157/3158	2.0	preS2 mRNA
2.1	3174/3175 5/6 113	1.95	S mRNAs
0.9[d]	1226, 1242	0.65–0.7	X mRNAs

[a] From Northern blots.
[b] Nucleotide numbering according to Galibert et al. (1979).
[c] Approximate length from 5′ end to site of polyA addition.
[d] Originally, four start sites were mapped by Treinin and Laub (1987); recent data show that there are only two (K. Yaginuma, personal communication).
Data from: Yaginuma et al. (1987); Will et al. (1987); Cattaneo et al. (1984); Standring et al. (1984)

Table 2. DHBV transcripts

Apparent size[a] (kb)	Start sites	Calculated length[b] (without polyA) (kb)	Function
3.5	2529	3.30	C mRNA; pregenome
2.4	732 ± 5 740 ± 5	2.05	preS mRNA
2.1	985 ± 2	1.80	S mRNA

[a] From Northern blots.
[b] Approximate length from 5′ end to site of polyA addition.
Data from Buscher et al. (1985)

structural components of the virion. The *genomic transcripts* comprise the mRNAs for synthesis of the proteins required to form replication-competent nucleocapsids and the RNA pregenome itself; the *subgenomic transcripts* serve as mRNAs for the expression of the preS2/S and S proteins which are major protein components of the viral envelope. In addition to these abundant transcripts, two relatively minor RNAs of 2.4 kb and 0.9 kb have been identified in HBV and correlated with the synthesis of a minor envelope protein termed preS1 (p 45; see chapter by D. GANEM, this volume) and of the X-gene product, respectively. Structural and functional properties of these transcripts are summarized in Table 1 (for DHBV in Table 2) and will be discussed below in more detail.

2.1 The Genomic Transcripts (3.5-kb RNAs)

As outlined in Fig. 1, these major RNAs comprise the entire genome length and include a short terminal redundancy which duplicates sequence elements that are essential for genome replication and circularization (see chapter by SEEGER et al., this volume). Their synthesis is liver-specifically controlled by the C-promoter (or genomic promoter) whose essential elements are confined within the 150 nucleotides preceeding the genomic RNA start site (Fig. 2; YAGINUMA and KOIKE 1989; YEE 1989; Sect. 5.1). This promoter directs the synthesis of two closely related, but functionally very different, subsets of genomic transcripts which, depending upon the presence or absence of the precore translation initiation site, have been designated according to their functions as precore mRNA and C-mRNA/pregenome, respectively (Fig. 2). Of these, the slightly shorter *C-mRNA/pregenome* is the most abundant. It is indispensable for viral genomic replication since it serves as mRNA for synthesis of core protein and, by internal translation initiation, of the P protein. In addition, after assembly with these gene products into core particles this RNA is the template for reverse transcription to produce the viral DNA genome. The 5′-end of the pregenomic RNA is unique and is located 5, 6 or 11 nucleotides downstream of the precore initiator codon in HBV, GSHV, and DHBV respectively (WILL et al. 1987; ENDERS et al. 1985; BÜSCHER et al. 1985).

The slightly longer *precore mRNAs* initiate 20–30 nucleotides upstream of the precore ORF (YAGINUMA et al. 1987), and serve as mRNAs for the synthesis of a fused precore/core gene product. As detailed in the chapter by D. GANEM, in this volume, this leads to the production of secreted and processed core gene products with structural and antigenic properties (e-antigen, HBeAg) quite different from the core protein assembled in the nucleocapsid (core antigen, HBcAg). Furthermore, translation of the precore ORF interferes with genomic RNA packaging and so the precore mRNAs are not packaged into nucleocapsids (NASSAL et al. 1990). Mutational analysis in the DHBV system (SCHLICHT et al. 1987; C. CHANG et al. 1987) and recent analysis of HBV genomes from chronically infected patients (BRUNETTO et al. 1990; OKAMOTO et al. 1990) have demonstrated that the precore/core gene products are not essential for hepadnaviral replication, and therefore the selective advantage of producing a genomic RNA encoding these products is not known.

2.2 The PreS2/S Transcripts (2.1-kb RNAs)

This second family of major RNAs is subgenomic in size and characterized by heterogeneous 5′ ends scattered throughout a region of about 150 nucleotides and bracketing the initiator for preS2/S gene translation (CATTANEO et al. 1983; STANDRING et al. 1984). In analogy to the genomic RNAs, these RNAs are templates for the synthesis of two variant gene products specified by the same reading frame. The several shorter RNAs (initiating between positions 1274 and 1395) serve for the expression of the major envelope protein (the small surface protein, p24). A longer major species (initiating at positions 1256/1257) allows translation of an envelope protein also containing the preS2 domain (the middle surface protein, p31). Again, synthesis of these several RNA species is apparently controlled by a single promoter element (the preS2/S promoter), which lacks a TATA transcriptional phasing signal. Instead, the proximal promoter elements located upstream of the CAP site of the preS2 mRNA (the most abundant 2.1-kb RNA species produced in vivo) display a strong similarity to GC-rich elements of the TATA-less SV40 late promoter (CATTANEO et al. 1983), the prototype promoter directing the synthesis of transcripts with heterogeneous 5′ ends. The preS2/S promoter is not stringently liver specific, and consequently, its transcripts and gene products are produced in substantial amounts also in nonliver cells (POURCEL et al. 1982).

2.3 The PreS1 Transcript (2.4-kb RNA)

The structure of this minor transcript suggests that it specifies the synthesis of the large surface protein (preS1, p45) which spans the preS1, preS2, and S domains (for details see chapter by D. GANEM, this volume). In contrast to most other HBV transcripts, the preS1 transcript is characterized by possessing a unique 5′ end which is probably positioned by a TATA motif mapping at the appropriate distance of 30 nucleotides upstream of the CAP site (WILL et al. 1987). The preS1 promoter element is highly liver specific, and is by far the least active HBV promoter element when tested in the isolated state with the help of a reporter gene (ANTONUCCI and RUTTER 1989; RANEY et al. 1990). Furthermore, preS1 transcripts have not so far been identified in RNA isolated from livers of WHV and GSHV infected animals (ENDERS et al. 1985; MÖRÖY et al. 1985).

2.4 The X Transcript

From its map position, a polyadenylated transcript of about 0.9-kb in size has become a prime candidate for a separate mRNA for X-gene expression. The production of this RNA species seems to be drastically repressed under conditions supporting viral replication since it has been detected neither in the HBV-infected liver, nor in hepatoma cells transfected with replication-competent viral DNA genomes. (As an exception of unknown significance, a nuclear polyA-transcript of 0.65-kb has been detected in the WHV-infected liver and been interpreted to be a precursor of the polyadenylated X-transcript, KANEKO and MILLER 1988.) The 0.9-kb transcript is, however, produced in substantial amounts from unconventional genomic constructs such as multiply repeated HBV genomes (GOUGH 1983) or HBV DNA contained in a lytic adenovirus vector (SAITO et al. 1985), or in the unusual environment of the testis of HBV-transgenic mice (ARAKI et al. 1989). Derepression of X-gene transcription is best observed when assayed on subgenomic transcriptional reading units starting about 200 bp upstream of the X transcription initiation site and containing the neighboring HBV enhancer element (enhancer I, see Sect. 5) as an essential component (TREININ and LAUB 1987). Thus, the enhancer I may be formally considered to be a part of the X promoter.

3 The PolyA Signal

All hepadnaviral RNAs terminate at a common position, about 20 nucleotides downstream of the conserved hexanucleotide TATAAA (AATAAA in DHBV) (GANEM and VARMUS 1987). An interesting feature of hepadnaviral transcription, shared by some other retroelements, is the ability to generate overlength transcripts by recognizing this signal only if it is located far from the 5′ end of the genomic precursor RNA (i.e., only after the transcription complex passes a second time through the polyadenylation signal).

Results from a recent deletion analysis in GSHV and HBV indicate that differential utilization of the polyA site depends on the presence of sequences located 5′ to this signal that act in *cis* to increase its use (RUSSNAK and GANEM 1990; PRIESS 1988). No such dependence on neighbouring sequences has been observed in initial experiments in a cell-free system, suggesting that *trans*-acting factors participate in the control of differential processing/polyadenylation (W. KELLER, personal communication).

4 Spliced Transcripts

According to our present knowledge, strategies of differential hepadnaviral gene expression rely on the large set of transcripts that are initiated at the four separate promoter regions. Hepadnaviruses thus do not use the alternative

Table 3. Spliced HBV RNAs derived from genomic transcripts

App. size[a] (kb)	Splice	Donor/Acceptor[b]	Calculated length[c] (without polyA) (kb)	Possible mRNA function
2.1	single	2447/489	2.05, 2.10	C-terminally truncated core proteins N-terminally truncated P protein
2.2	double	2067/2350; 2447/282	2.00	Fusion of internally deleted C protein to N-terminally truncated S protein N-terminally truncated P protein

[a] From Northern blots.
[b] Nucleotide numbering according to GALIBERT et al. (1979).
[c] Approximate length from 5′ end to site of polyA addition.
Data from: SUZUKI et al. (1989), SU et al. (1989), CHEN et al. (1989).

possibility of activating internal coding sequences by differential splicing, as occurs in the related retrovirus family, which produce all of their subgenomic mRNAs by splicing of a single primary transcript (the genomic RNA/*gag* mRNA). In lentiviruses multiple RNA splicing is used to create fused gene products with novel properties, including the regulation of stage-specific viral gene expression (CULLEN and GREENE 1990). In principle, such a scheme for optimizing functional coding capacity from limited sequence information appears to be particularly well suited for the hepadnaviruses. Therefore, numerous attempts have been made to identify spliced transcripts in HBV-infected tissue and to correlate these with viral replication functions. Initial evidence for potential splice sites in HBV RNA was obtained in transfection experiments with chimeric SV40/HBV genomes (SIMONSEN and LEVINSON 1983; CATTANEO et al. 1983). Very recently, however, several groups have also identified two spliced transcripts (Table 3) in cells transfected with replication-competent HBV genomes, and in the livers of chronically infected patients (CHEN et al. 1989; SUZUKI et al. 1989; SU et al. 1989). Similarly, low levels of spliced transcripts were detected in the animal virus models (W. TUCKER, unpublished observations; D. GANEM, personal communication). In no case tested, however, has mutational inactivation of splice donor or acceptor sequences resulted in a detectable loss of competence for viral replication. Thus, it presently remains unclear whether the low level of RNA splicing detected is of any functional significance in hepadnaviral gene expression.

5 Regulation of Viral Transcription

5.1 *Cis*-Acting Elements Regulating Transcriptional Activity

Besides the four functionally defined promoters and the single polyadenylation site, two enhancer regions with liver-specific characteristics have been detected in the HBV genome. One is located in the upstream parts of the X promoter (generally known as *the* enhancer, SHAUL et al. 1985; TOGNONI et al. 1985), very recently designated enhancer I, and the other lies upstream of the C promoter (designated liver-specific enhancer, LSE, YEE 1989; or enhancer II). These regions have been shown to stimulate the activity of distally located promoters when removed from their natural context. In addition, a glucocorticoid responsive element (GRE; TUR-KASPA et al. 1986, 1988) has been shown to stimulate S-gene expression and also to exert its function probably in conjunction with enhancer I.

All promoter elements and their constituent sequence motifs have been studied in great detail by conventional techniques, primarily by linking each promoter element in isolation to a reporter gene such as chloramphenicol acetyltransferase (*CAT*), and then testing these constructs for gene expression after gene transfer into cultured cells. Using this approach, signal strength and species specificity of these elements have been monitored in cell lines of various origin. In addition, the size limits required for maximal promoter activity have been determined by deletion analysis (for a recent comprehensive review, see SHAUL 1991). As defined by this experimental system, promoter elements were found to be rather short and to range from 100 to 300 bp in length (Fig. 2). In HepG2 cells, their strength was found to decrease in the order $C > X > S2/S > S1$ (ANTONUCCI and RUTTER 1989); in linearized genomes their order was $X = S2/S > C \gg S1$ (RANEY et al. 1990). Both orders differ significantly from the steady-state ratio of transcripts observed in vivo ($S > C \gg S1 \geqslant X$; WILL et al. 1987). Only the core promoter and the preS promoter appear to display stringent tissue specificity. The activity of these promoters was significantly increased above basal levels when these elements were tested coupled to DNA segments containing the enhancer I element (L.J. CHANG et al. 1990; ANTONUCCI and RUTTER 1989; HONIGWACHS et al. 1989), and the tissue specificity of the S promoter was also markedly enhanced. In contrast, the X promoter displayed considerably less tissue specificity although it contains the same enhancer element (RANEY et al. 1990; ANTONUCCI and RUTTER 1989). Furthermore, a negative influence of preS2/S promoter sequences on preS1 promoter activity has been observed by BULLA and SIDDIQUI (1988).

The deletion mapping of the promoter/enhancer elements has been complemented by an extensive mapping of discrete nucleotide sequences which are defined by their interaction with sequence specific DNA binding factors. As listed in Table 4, and depicted in Fig. 3 in detail for the core gene promoter and for the enhancer/X promoter from HBV, these elements have the potential to bind stably to a multitude of nuclear protein factors as detected by "footprinting" and "mobility shift" in cell-free assay systems. Some of the factors binding to HBV

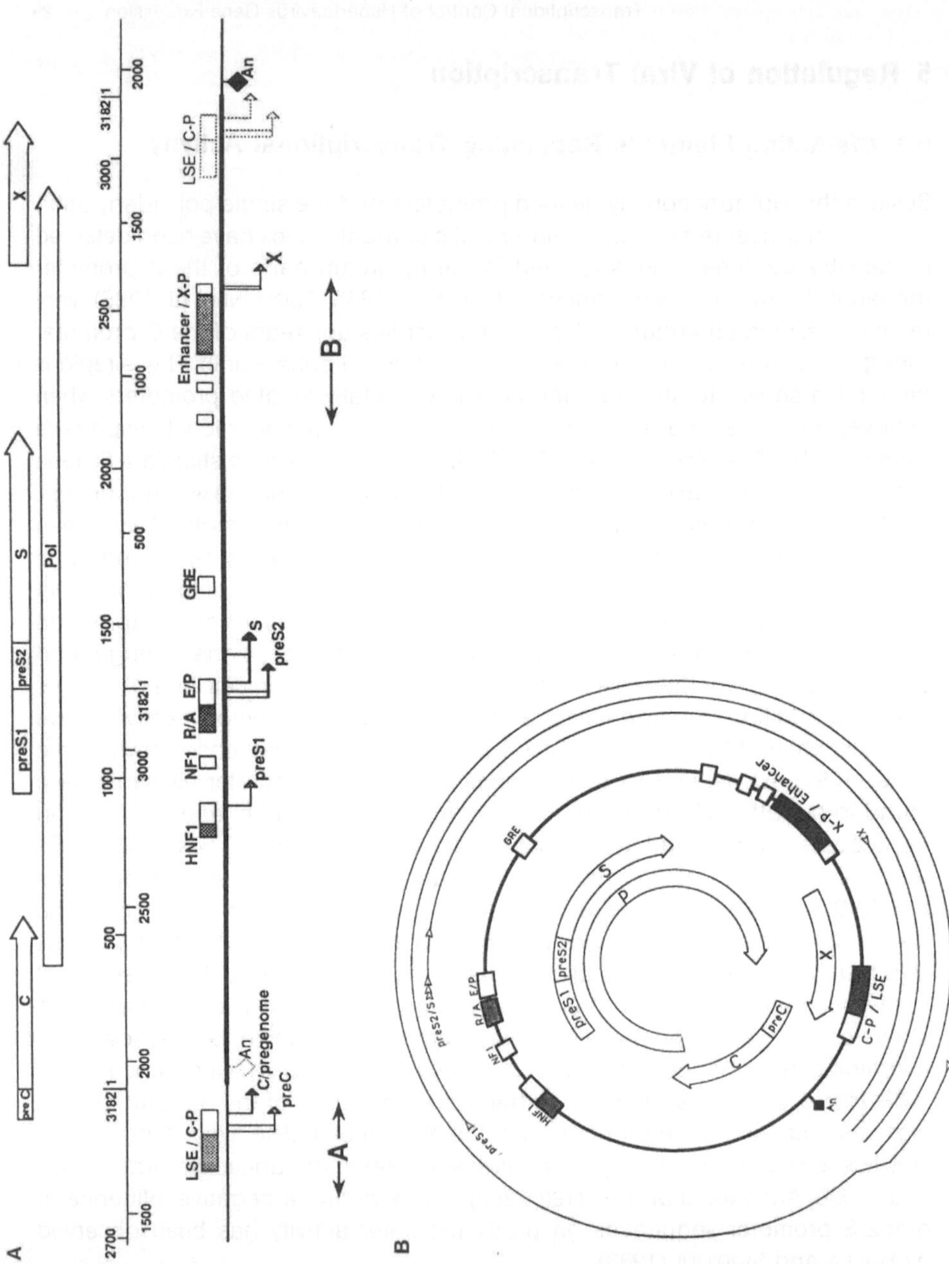

Fig. 2 A, B. Localization of transcriptional control elements in the HBV genome. **A** Linear representation with terminal redundancy duplicating regions involved in control of genomic RNA synthesis. *Open arrows* represent open reading frames. Nucleotide numbering according to Pasek et al. (1979) (*above* the line) and Galibert et al. (1979) (*below* the line). For abbreviations of control elements, see Table 4. Control elements are drawn as *boxes*, elements exhibiting liver specificity are *shaded*. Transcriptional start sites are indicated by *small arrows*. A_n, polyadenylation signal. Control elements and signals are shown in **boldface** when functionally active in the displayed position. *Lighter print* indicates inactive duplications. Protein binding sites in the areas marked *A* and *B* are shown in greater detail in Fig. 3 *A, B*. **B** Circular representation facilitating comparison with Fig.1

Table 4. Transcriptional control elements in the HBV genome

Site	Position	Comment
PreS1 Promoter region		
HNF1	2717–2739	HNF1 binding site. Also identified as AFP1 binding site [1, 2]
PreS2/S Promoter region		
NF1	2967–2986	NF1 binding site [3]
R/A	3079–3134	Sequence activates preS/S promoter in liver cells, but represses in HeLa cells (4)
E/P	3155–30	Enhancer/promoter-like element, nonspecific, suppressed by adjacent negative regulator R/A. Homologies to API, CRE, GCN4 binding sites and to SRE (LDL sterol response element) [4]
GRE	341–370	Glucocorticoid receptor binding site, biologically functional in conjunction with HBV enhancer [5]
Enhancer/X Promoter Region (Enh/X-P)		
UE3	882–901	Footprint of an unidentified protein [6]
FP VIII	977–994	Binding site of the liver-specific C/EBP [7], probably identical to UE2 (973–992) [6]
FP VII	998–1004	NFkB binding site [7]
FP VI	1026–1050	Binding site of the liver-specific C/EBP and HNF1 [8]. Coincides with UE1 (1030–1049) [6], also binds OCT2 [7]
FP IV	1101–1119	NF1 binding site, also mapped as NF1c by Ben-Levy et al. [9] to positions 1097–1114
FP V	1128–1139	Binding site of a liver-specific protein factor [8]
FP III	1153–1172	EFC binding site [8]. Also identified as EP (1145–1170) by Shaul et al. [6]
eH-TF		Binding site of eH-TF [7], also identified as TGT3 element by Ben–Levy et al. [9], overlaps FP III
NF1	1159–1189	NF1 binding site (NF1b in Ben–Levy et al. [9], overlaps FP II and III
FP II	1174–1199	C/EBP [8] and AP1 [10] binding site, cAMP induction may be mediated by binding of CREB to a CRE homology in this element [10], coincides with the element E mapped by Shaul et al. [6]
FP 1	1205–1235	NF1 binding site [8]
C-Promoter Region (C-P/LSE)		
FP I	1599–1615	Footprints induced by rat liver nuclear extracts [11]
FP II	1619–1632	
FP III	1637–1703	Footprints containing two binding sites of the liver-specific factor EBP binding sites [11]
FP IV	1708–1743	Liver-specific footprint [11], also identified as 3.6 PBP (1715–1731) by Yaginuma et al. [12], possibly a binding site of eH-TF [7]
FP V	1749–1777	Liver-specific footprints containing several EBP binding sites [11]
FP VI	1801–1827	
FP VII	1833–1850	
LSE	1685–1773	Liver-specific element (enhancer II) [13]

[a] Nucleotide numbering according to Galibert et al. (1979).
Data from: [1] Nakao et al. (1989); [2] Courtois et al. (1988); [3] Shaul et al. (1986); [4] De Medinal et al. (1988); [5] Tur Kaspa et al. (1986, 1988); [6] Shaul and Ben–Levy (1987); [7] A. Siddiqui, personal communications; [8] Patel et al. (1989); [9] Ben-Levy et al. (1989); [10] Faktor et al. (1990); [11] Lopez-Cabrera et al. (1990); [12] Yaginuma and Koike et al. (1989); [13] Yee (1989); for review see also Shaul (1991)

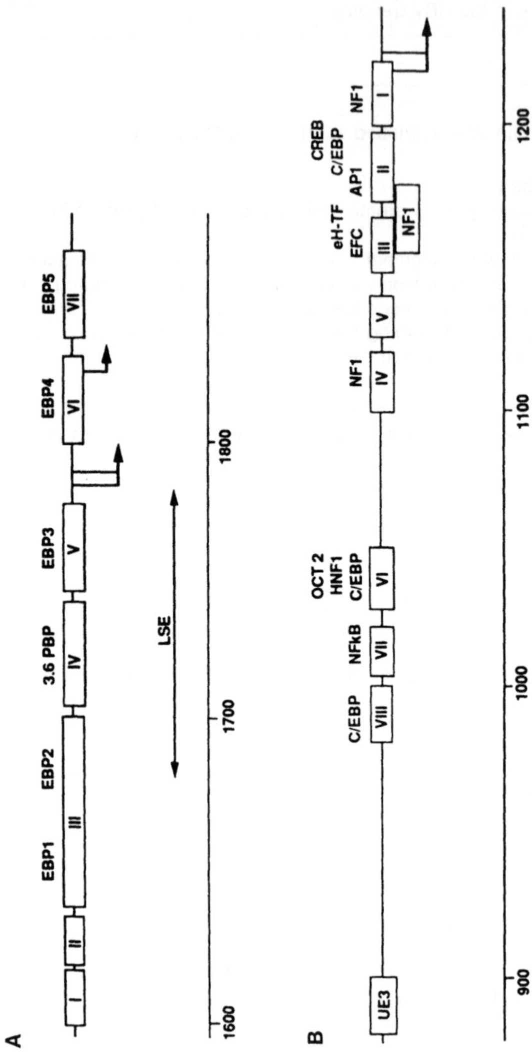

Fig. 3 A, B. Protein binding sites in the two main regulatory regions of the HBV genome. **A** C promoter/LSE. Footprints are numbered *I–VII* according to Lopez-Cabrera et al. (1990). For abbreviations, see Table 4. Nucleotide numbering according to Galibert et al. (1979). **B** Enhancer/X promoter. Footprints are numbered *I–VII* according to Patel et al. (1989). For abbreviations, see Table 4. Nucleotide numbering according to Galibert et al. (1979)

DNA sequences (NF1, C/EBP, HNF1, AP1) are structurally and functionally well defined and have also become available in pure form by recombinant DNA methods. The majority of factors, however, are defined primarily by the nucleotide sequences which they recognize (see also SHAUL 1991).

Altogether, these data reveal a dense clustering of protein binding sites along the promoter regions analyzed. It should, however, be noted that these clusters extend beyond the limits required for promoter function as defined above (Fig. 3), and therefore many more protein bindings sites may be located outside of the promoter sequences that have so far been preferentially investigated. In addition, it should be kept in mind that the protein interaction sites detected by the techniques presently used may only display the stable and less flexible part of DNA/protein and protein/protein interactions. Considering that the distance between any two sequence elements in the viral DNA genome is short relative to the distances of 3–10 kb often observed for single cellular promoter/enhancer regions, the whole 3-kb HBV genome should be considered in toto as a single, complex "regulon" in which regulatory proteins binding to any given *cis*-acting element may productively interact, directly or indirectly, with several other such sites and exert combinatorial effects not detectable with subgenomic DNA elements.

5.2 Regulation of Transcription by Viral Gene Products

Recently it has been shown that the X-gene product is capable of stimulating the activity of a variety of viral and cellular promoters including that of the HBV enhancer and the core gene promoter or the S gene promoter (Twu and Schloemer 1987; Spandau and Lee 1988; Zahm and Hofscheider 1988; Colgrove et al. 1989; Twu and Robinson 1988; Aufiero and Schneider 1990; see Schek et al. 1990 for a recent review). The mechanism of this interaction is presently unclear. But interaction of the X-protein with its many targets seems unlikely in view of the very low X-protein concentration in the HBV-infected liver (Pfaff et al. 1987). Instead, direct or indirect modulation of factors regulating viral and cellular gene activity appears to be a more realistic alternative to explain the pleiotropic transactivation properties of the X-gene product. In keeping with this notion, very recent data suggest that the X-protein may possess protein kinase activity (Wu et al. 1990), or belong to a family of protease inhibitors (T. Tanaka and K. Koike, personal communication). A 3- to 10-fold stimulation of the synthesis of HBsAg and HBcAg and of genomic transcripts by X-transactivation has been observed in differentiated human hepatoma cells during transient expression of HBV genomes lacking a functional X-gene (unpublished results: Colgrove et al. 1989), and similarly increased levels of activity have been reported for the corresponding promoters if coupled individually to indicator genes such as *CAT* or *LUC* (Colgrove et al. 1989, Raney et al. 1990). In the latter experimental setting, strong transactivation was also observed on the enhancer/X promoter element (Faktor and Shaul 1990). Extrapolated to the HBV genome, this suggests that the X-gene product could stimulate its own synthesis. If so, this would obviously require a mechanism for downregulation of X expression after a desired level of X protein has been reached. In agreement with such an autoregulatory scheme, a reduction of HBV gene expression has been observed at high levels of X-gene dosage in transfection experiments (Faktor and Shaul 1990; unpublished data from this laboratory).

Several recent reports suggest that viral gene products encoded outside of the X-gene may also participate in the transcriptional control of viral gene

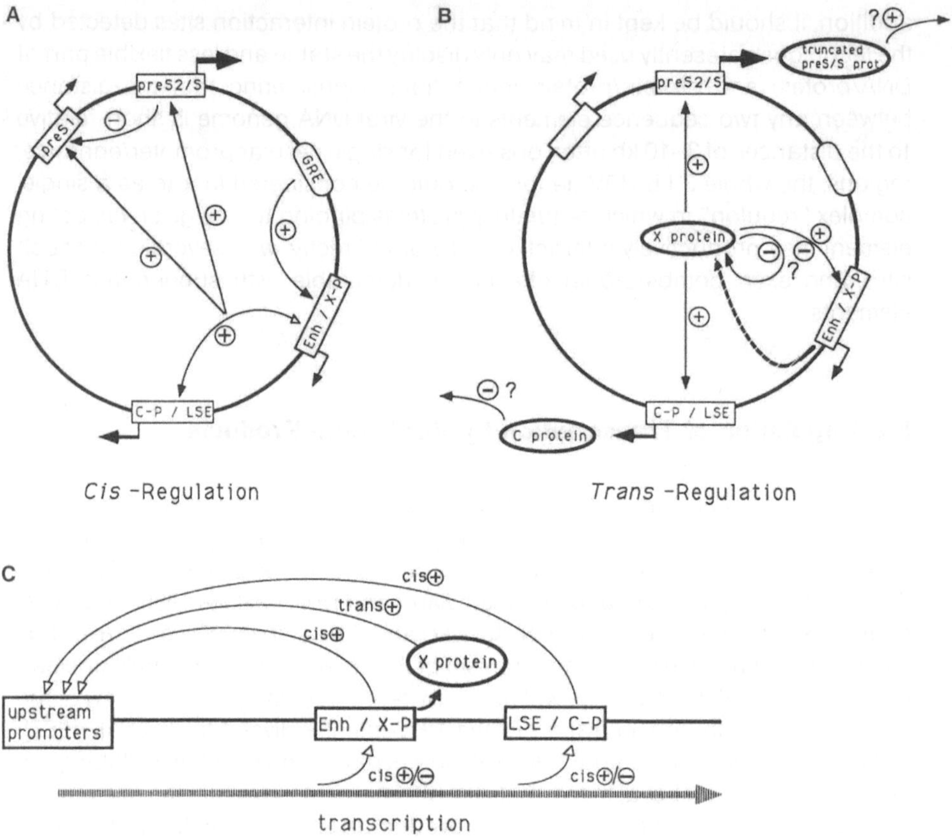

Fig. 4 A–C. Regulatory interactions governing HBV transcription. **A** Interactions of *cis* regulatory elements. **B** Proteins and targets involved in transregulation. **C** Hypothetical autoregulatory scheme involving the two transcribed enhancer/promoter regions and the X transactivator. See Sect. 5.3

expression. A 3′-truncated preS2/S region of integrated HBV DNA expressed in a human hepatoma cell line and tissue was found to encode a transcriptional transactivator (KEKULE et al. 1990), suggesting that additional routes may exist for the generation of transacting functions by HBV DNA integration. Negative regulation of a cellular gene by the core gene product has been reported by TWU and SCHLOEMER (1989) and the same group has also observed downregulation of an S promoter/enhancer-driven *CAT* gene (personal communication). Finally, a regulatory function of a core gene product is also suggested by the observation that core proteins are targeted to the nucleus due to possessing an intrinsic nuclear targeting signal (OU et al. 1989).

5.3 Modulation of Transcriptional Activity by Transcription Through *cis*-Regulatory DNA Elements?

As previously noted, all genomic signals lie within sequences actively transcribed during viral replication and are probably evolutionarily adapted to this situation. In keeping with this notion, activation of transcription by enhancer I was found to be much more pronounced if the enhancer element was located within, instead of upstream of the transcribed sequence (VANNICE and LEVINSON 1988). Similarly, transcribed enhancer elements could potentially also function to monitor and to adjust in *cis* the overall activity of upstream promoters directing these transcripts, in analogy to the downregulation of promoter activity by transcription from upstream as observed with serially arranged promoter elements (promoter occlusion; ADHYA and GOTTESMAN 1982). In the circular HBV genome the central control region consisting of the X gene plus two flanking *cis* regulatory regions is located downstream of the promoters activated by these regions (Fig. 2B, 4C). Thus, modulation of upstream promoter activity by transcriptionally induced changes in the downstream promoter/enhancer elements could either be transmitted in *cis* by enhancer/promoter interactions or, alternatively, in *trans* involving the X-gene product whose production is controlled by the X promoter/enhancer I (Fig. 4C).

5.4 Evolution of the Region Controlling Genomic RNA Synthesis: Two Enhancer/Promoter Regions Derived from a Short Continuous Element by X-Gene Insertion?

As discussed above, for HBV the activity of the genomic RNA promoter is controlled by its coordinate interaction with the distally located enhancer/X promoter, and additionally modulated by the action of the X-gene product encoded between these two elements. In DHBV, as already suggested by the absence of the X-gene, and as confirmed by deletion analysis, this extended central control region is replaced by a short continuous enhancer/promoter region (Fig. 5A). This region of about 350 bp has been further dissected into a basal promoter element (positions 2351–2531) and an adjacent upstream sequence (positions 2172–2351) fulfilling the requirements of a classical enhancer (C. LIU and W. MASON, personal communication). Furthermore, nuclear protein factors from duck liver have been shown to bind to two DNA fragments (positions 2401–2489 and 2202–2296) mapping to the basal element and the enhancer of the DHBV genomic promoter (FISCHER 1987; Fig. 5B). Both of these protein binding regions were found to contain significant sequence similarities to important protein binding sites in the HBV enhancer/X promoter. The avian hepadnavirus genome thus appears to be more closely related to that of a putative common ancestor from which the more complex mammalian viruses may have arisen after integration, probably from the host genome (MILLER and ROBINSON 1986), of an additional regulatory gene into their

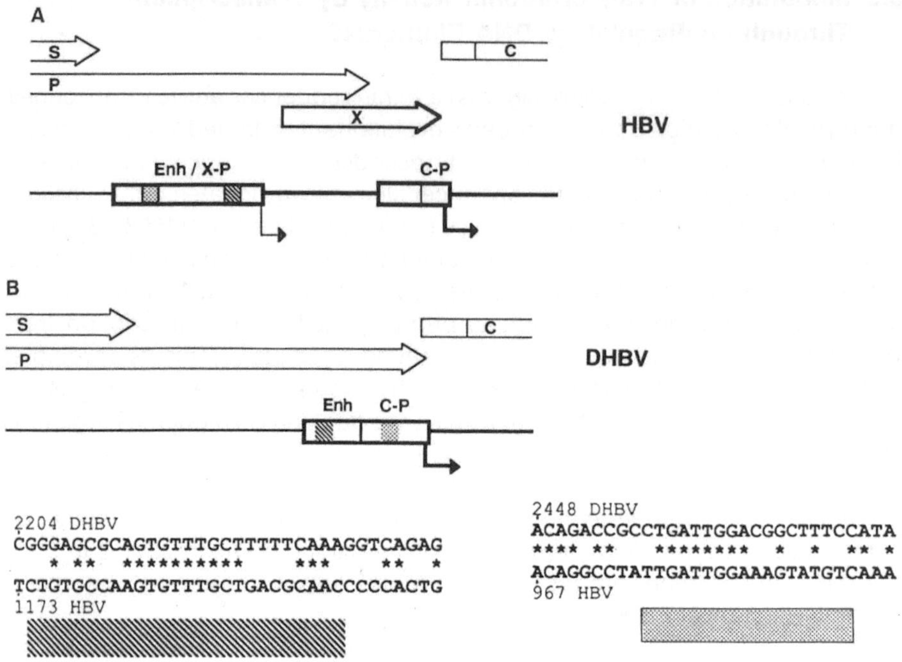

Fig. 5A, B. Regulatory regions controlling genomic RNA synthesis in mammalian (HBV) and avian (DHBV) hepadnaviruses. **A** Comparative schematic representation. *Stippled box*, FP VIII (see Table 4) in HBV and homologous sequence in DHBV. *Hatched box*, area around FP II (see Table 4) and homologous sequence in DHBV. **B** Comparison of the respective sequences. Nucleotides conserved between the two viral genomic sequences are indicated by *asterisks*

"conventionally" organized genomic enhancer/promoter region. Consecutive adaptational changes may then have resulted in the present extended control region consisting of two largely independent enhancer/promoter regions flanking the X-gene.

6 Conclusions and Perspectives

Recent years have seen major advances in our knowledge of both hepadnaviral transcription and the molecular mechanisms involved in its regulation. Although the major contributors have been identified, the details of interactions between cellular and viral factors are far from being understood. Outstanding questions concern the pattern of long range interactions between control elements in the transcriptionally active viral DNA and the nature of the resulting regulatory circuits. Still less is currently known about the mechanism(s) of X protein action and how its synthesis and activity are regulated. As previously noted, there may exist as yet unknown viral gene products acting directly or indirectly on viral transcription, as well as spliced RNA species which could open new means of regulating gene expression. To address these questions it may be necessary for future experimentation to abandon some of the simplifying approaches currently used and to aim increasingly at simulation of events within hepatocytes infected with intact viral genomes. Thus, the effects of mutational changes should also be studied in the context of the circular viral genome. Furthermore, low copy numbers of viral genomes per cell should be employed to avoid unbalanced gene expression or titration of cellular proteins. Infection rather than transfection, including the use of defective viruses, may be needed to meet these requirements. Considering the presently available experimental systems, emphasis may shift at least transiently to animal virus models. Of these, the DHBV system may be particularly attractive since it appears to be the simplest and evolutionarily least developed. Comparative studies involving both the human and the avian viruses may help to understand the fundamental pathways governing hepadnaviral transcription.

Acknowledgements. We thank our many colleagues for providing reprints and preprints, Laura Runkel for constructive criticism, and also Karin Coutinho for secretarial assistance. The author's work cited here was supported by the Deutsche Forschungsgemeinschaft and the Fond der Chemischen Industrie.

References

Adhya S, Gottesman M (1982) Promoter occlusion transcription through a promoter may inhibit its activity Cell 29: 939–944

Antonucci TK, Rutter WJ (1989) Hepatitis B virus (HBV) promoters are regulated by the HBV enhancer in a tissue-specific manner. J Virol 63: 579–583

Araki K, Miyazaki JI, Hino O, Tomita N, Chisaka O, Matsubara K, Yamamura KI (1989) Expression and replication of hepatitis B virus genome in transgenic mice. Proc Natl Acad Sci USA 86: 207–211

Aufiero B, Schneider RJ (1990) The hepatitis B virus X-gene product transactivates both RNA polymerase II and III promoters. EMBO J 9: 497–504

Ben-Levy R, Faktor O, Berger I, Shaul Y (1989) Cellular factors that interact with the hepatitis B virus enhancer. Mol Cell Biol 9: 1804–1809

Brunetto MR, Stemler M, Bonino F, Schödel F, Oliveri F, Rizzetto M, Verme G, Will H (1990) A new hepatitis B virus strain in patients with severe anti-HBc positive chronic hepatitis. J Hepatol 10:258–261

Bulla G, Siidiqui A (1988) The hepatitis B virus enhancer modulates transcription of the hepatitis B virus surface antigen gene from an internal location. J Virol 62: 1437–1441

Büscher M, Reiser W, Will H, Schaller-H (1985) Transcripts and the putative RNA pregenome of duck hepatitis B virus implications for reverse transcription. Cell 40: 717–724

Cattaneo R, Will H, Hernandez N, Schaller H (1983) Signals regulating hepatitis B surface antigen transcription. Nature 305: 336–338

Cattaneo R, Will H, Schaller H (1984) Hepatitis B virus transcription in the infected liver. EMBO J 13: 2191–2196

Chang C, Enders G, Sprengel R, Peters N, Varmus HE, Ganem D (1987) Expression of the precore region of an avian hepatitis B virus is not required for viral replication. J Virol 61: 3322–3325

Chang LJ, Ganem D, Varmus HE (1990) Mechanism of translation of the hepadnaviral polymerase (P) gene. Proc Natl Acad Sci USA 87: 5158–5162

Chen PJ, Chen CR, Sung JL, Chen DS (1989) Identification of a doubly spliced viral transcript joining the separated domains for putative protease and reverse transcriptase of hepatitis B virus. J Virol 63: 4165–4171

Colgrove R, Simon G, Ganem D (1989) Transcriptional of homologous and heterologous genes by the hepadnaviral X gene product in cells permissive for viral replication. J Virol 63: 4019–4026

Courtois G, Baumhueter S, Crabtree GR (1988) Purified hepatocyte nuclear factor 1 interacts with a family of hepatocyte specific promoters. Proc Natl Acad Sci USA 85: 7937–7941

Cullen BR, Greene WC (1990) Regulatory pathways governing HIV-1 replication. Cell 58: 423–426
De Medina T, Faktor O, Shaul Y (1988) The S promoter of hepatitis B virus is regulated by positive and negative elements. Mol Cell Biol 8: 2449–2455
Enders GH, Ganem D, Varmus H (1985) Mapping the major transcripts of ground squirrel hepatitis virus the presumptive template for reverse transcriptase is terminally redundant. Cell 42: 297–308
Faktor O, Shaul Y (1990) The identifcation of hepatitis B virus X gene responsive elements reveals functional similarity of X and HTLV-I tax. Oncogene 6: 867–872
Faktor O, Budlovsky S, Ben-Levy R, Shaul Y (1990) A single element within the hepatitis B virus enhancer binds multiple proteins and responds to multiple stimuli. J Virol 64: 1861–1863
Fischer M (1987) Protein/DNA Wechselwirkungen am Genom des Enten Hepatitis B Virus (DHBV). Diploma thesis (Diplomarbeit), University of Heidelberg
Galibert F, Mandart E, Fitoussi F, Tiollais P, Charnay P (1979) Nucleotide sequence of the hepatitis B virus genome (subtype ayw) cloned in *E. coli.* Nature 281: 646–650
Ganem D, Varmus HE (1987) The molecular biology of hepatitis B viruses. Annu Rev Biochem 56: 651–693
Gough N (1983) Core and e antigen synthesis in rodent cells transformed with hepatitis B virus DNA is associated with greater than genome length viral messenger RNAs. J Mol Biol 165: 683–699
Honigwachs J, Faktor O, Dikstein R, Shaul Y, Laub O (1989) Liver-specific expression of hepatitis B virus is determined by the combined action of the core gene promoter and the enhancer. J Virol 63: 919–924
Kaneko S, Miller RH (1988) X-region specific transcription in mammalian hepatitis B virus infected liver. J Virol 62: 3979–3984
Kekule AS, Lauer U, Meyer M, Caselmann WH, Hofschneider PH, Koshy R (1990) The pres2/s region of integrated hepatitis B virus DNA encodes a transcriptional transactivator. Nature 343: 457–461
Landschulz W, Johnson PF, Adashi EY, Graves B, McKnight SL (1988) Isolation of a recombinant copy of the gene encoding C/EBP. Genes Dev 2: 786–800
Lopez–Cabrera M, Letovsky J, Hu KQ, Siddiqui A (1990) Multiple liver-specific factors bind to the hepatitis B virus core/pregenomic *Trans*-activation and repression by CAAT/enhancer binding protein. Proc Natl Acad Sci USA 87: 5069–5073
Miller RH, Robinson WS (1986) Common evolutionary origin of hepatitis B viruses and retroviruses. Proc Natl Acad Sci USA 83: 2531–2535
Möröy J, Etiemble J, Trepo C, Tiollais P, Buendia MA (1985) Transcription of woodchuck hepatitis virus in the chronically infected liver. EMBO J 4: 1507–1514
Nakao K, Miyao Y, Ohe Y, Tamaoki T (1989) Involvement of an AFP-1 binding site in cell specific transcription of the pre-SI region of the human hepatitis B virus surface antigen gene. Nucleic Acids Res 17: 9833–9842
Nassal M, Junker–Niepmann M, Schaller H (1990) Translation inactivation of RNA function Discrimination against a subset of genomic transcripts during hepatitis B virus nucleocapsid assembly. Cell 63: 1357–1363
Okamoto H, Yotsumoto S, Akahane Y, Yamanaka T, Miyazaki Y, Sugai Y, Tsuda F, Tanaka T, Miyakawa Y, Mayumi M (1990) Hepatitis B viruses with precore region defects prevail in persistently infected hosts along with seroconversion to the antibody against e antigen, J Virol 64: 1298–1303
Ou JH, Yeh CT, Yen TS (1989) Transport of hepatitis B virus precore protein into the nucleus after cleavage of its signal peptide. J Virol 63: 5238–543
Pasek M, Goto T, Gilbert W, Zink B, Schaller H, McKay P, Leadbetter G, Murray K (1979) Hepatitis B virus genes and their expression in *E.* coli. Nature 282: 575–579
Patel NU, Jameel S, Isom H, Siddiqui A (1989) Interactions between nuclear factors and the hepatitis B virus enhancer. J Virol 63: 5293–5301
Pfaff E, Salfeld J, Gmelin K, Schaller H, Theilmann L (1987) Synthesis of the X protein of hepatitis B virus in vitro and detection of anti X antibodies in human sera. Virology 158: 456–460
Pourcel C, Louise A, Gervais M, Chenciner N, Dubois MF, Tiollais P (1982) Transcription of the hepatitis B surface antigen gene in mouse cells transformed with cloned viral DNA. J virol 42: 100–105
Priess A (1988) System zur Untersuchung spezifischer Aspekte der 3′-Prozessierung von HBV mRNAs. Diploma thesis (Diplomarbeit), University of Heidelberg
Raney AK, Milich DR, Easton AJ, McLachlan A (1990) Differentiation-specific transcriptional regulation of the hepatitis B virus large surface antigen gene in human hepatoma cell lines. J Virol 64: 2360–2368
Russnak R, Ganem D (1990) Sequences 5′ to the polyadenylation signal mediate differential poly(A) site use in hepatitis B viruses. Genes Dev 4: 764–776

Saito I, Oya Y, Yamamoto K, Yuasa T, Shimojo H (1985) Construction of nondefective adenovirus type 5 bearing a 2.8-kilobase hepatitis B virus DNA near the right end of its genome. J Virol 54: 711–719

Schek N, Fischer M, Schaller H (1990) The hepadnaviral X protein in, In: McLachlan A (ed) Molecular biology of hepatitis B viruses, CRC, Boca Raton (CRC Uniscience)

Schlicht, HJ, Salfeld J, Schaller H (1987) The preC region of the duck hepatitis B virus is essential to synthesis and secretion of processed core proteins but not for virus formation J Virol 61: 3701–3709

Shaul Y (1991) Regulation of hepadnavirus transcription. In: McLachlan A (ed) Molecular biology of hepatitis B viruses. CRC, Boca Raton (CRC Uniscience)

Shaul Y, Ben–Levy R (1987) Multiple nuclear proteins in liver cells are bound to hepatitis B virus enhancer element and its upstream sequences. EMBO J 6: 1913–1920

Shaul Y, Rutter WJ, Laub O (1985) A human hepatitis B viral enhancer element. EMBO J 4: 427–430

Shaul Y, Ben-Levy R, De Medina T (1986) High affinity binding site for nuclear factor I next to the hepatitis B virus S gene promoter. EMBO J 5: 1967–1971

Simonsen CC, Levinson AD (1983) Analysis of processing and polyadenylation signals of the hepatitis B virus surface antigen gene by using simian virus 40–hepatitis B virus chimeric plasmids. Mol Cell Biol 3: 2250–2258

Spandau DF, Lee CH (1988) *Trans*-activation of viral enhancers by the hepatitis B virus X protein. J Virol 62: 427–434

Standring DN, Rutter WJ, Varmus HE, Ganem-D (1984) Transcription of the hepatitis B surface antigen gene in cultured murine cells initiates within the presurface region. J Virol 50: 563–571

Su TS, Lai CJ, Huang JL, Lin LH, Yauk YK, Chang CM, Lo SJ, Han SH (1989) Hepatitis B virus transcript produced by RNA splicing. J Virol 63: 4011–4018

Suzuki T, Masui N, Kajino K, Saito I, Miyamura T (1989) Detection and mapping of spliced RNA from a human hepatoma cell line transfected with the hepatitis B virus genome. Proc Natl Acad Sci USA 86: 8422–8426

Tognoni A, Cattaneo R, Serfling E, Schaffner W (1985) A novel expression selection approach allows precise mapping of the hepatitis B virus enhancer. Nucleic Acids Res 13: 7457–7472

Treinin M, Laub O (1987) Identification of a promoter element located upstream from the hepatitis B virus X gene. Mol Cell Biol 7: 545–548

Tur-Kaspa R, Burk RD, Shaul Y, Shafritz DA (1986) Hepatitis B virus DNA contains a glucocorticoid-responsive element. Proc Natl Acad Sci USA 83: 1627–1631

Tur-Kaspa R, Shaul Y, Moore DD, Burk RD, Okret S, Poellinger L, Shafritz DA (1988) The glucocorticoid receptor recognizes a specific nucleotide sequence in hepatitis B virus DNA causing increased activity of the HBV enhancer. Virology 167: 630–633

Twu JS, Robinson WS (1988) Hepatitis B virus X gene can transactivate heterologous viral sequences. Proc Natl Acad Sci USA 86: 2046–2050

Twu JS, Schloemer RH (1987) Transcriptional transactivating function of hepatitis B virus. J Virol 61: 3448–3453

Twu JS, Schloemer RH (1989) Transcription of the human beta interferon gene is inhibited by hepatitis B virus. J Virol 63: 3065–3071

Vannice JL, Levinson AD (1988) Properties of the human hepatitis B virus enhancer position effects and cell-type nonspecificity. J Virol 62: 1305–1313

Wickens M (1990) How the messenger got its tail addition of poly(A) in the nucleus. Trends Biochem Sci 15: 277–281

Will H, Reiser W, Weimer T, Pfaff E, Büscher M, Sprengel R, Cattaneo C, Schaller H (1987) Replication strategy of human hepatitis B virus. J Virol 61:904–911

Wu JY, Zhou ZY, Judd A, Cartwright CA, Robinson WS (1990) The hepatitis B virus encoded transcriptional *trans*-activator hbx appears to be a novel protein serine/threonine kinase. Cell 63: 687–695

Yaginuma K, Koike K (1989) Identification of a promoter region for 3.6 kilobase mRNA of hepatitis B virus and specific cellular binding protein. J Virol 63: 2914–2921

Yaginuma K, Shirakata Y. Kobayashi M, Koike K (1987) Hepatitis B virus (HBV) particles are produced in a cell culture system by transient expression of transfected HBV DNA. Proc Natl Acad Sci USA 84: 2678–2686

Yee JK (1989) A liver-specific enhancer in the core promoter region of human hepatitis B virus. Science 246: 658–661

Zahm P, Hofschneider PH, Koshy R (1988) The HBV X-ORF encodes a transactivator, a potential factor in viral hepatocarcinogenesis. Oncogene 3: 169–177

Viral DNA Synthesis

C. SEEGER,[1] J. SUMMERS,[2] and W. S. MASON[3]

1 Introduction

In 1973, it was first discovered that human hepatitis B virus (HBV) preparations from sera of chronically infected individuals contain an endogenous polymerase activity capable of carrying out a repair reaction on the viral DNA template in the presence of magnesium and nucleotide triphosphates (KAPLAN et al. 1973). The polymerase was shown to be resistant to solubilization by nonionic detergents, a condition precluding the characterization of enzymatic activities with exogenous RNA or DNA templates. With the help of electron microscopic characterization of

[1] Department of Microbiology, Immunology and Parasitology, New York State College of Veterinary Medicine, Cornell University, Ithaca, NY 14853, USA
[2] Department of Cell Biology, UNM Cancer Center, University of New Mexico, School of Medicine, Albuquerque, NM 871311, USA
[3] Institute for Cancer Research, Fox Chase Cancer Center, 7701 Burholme Ave., Philadelphia, PA 19111, USA

Current Topics in Microbiology and Immunology, Vol. 168

the predominant species in virus preparations (Robinson et al. 1974), together with sedimentation and electrophoretic analysis of the DNA products of the endogenous polymerase reaction (Kaplan et al. 1973; Summers et al. 1975), the structure of the viral genome was identified. HBV has a ca. 3 kbp-long genome held in a relaxed circular conformation by a short, cohesive overlap (ca. 0.2 kbp) between the 5′ ends of the two DNA strands (Summers et al. 1975; Sattler and Robinson 1979; Ganem et al. 1982; Molnar-Kimber et al. 1984). One strand, subsequently shown to encode the viral proteins and hence termed minus-strand DNA (Pasek et al. 1979), is complete whereas the other strand, plus-strand DNA, is incomplete and heterogeneous in length (Summers et al. 1975). Thus, the repair reaction carried out by the endogenous polymerase in vitro fills in the single-stranded gap, which can span up to 50% of the viral genome. These characteristics of the HBV genome have served, along with other criteria, to identify additional members of the hepadnaviral family, including woodchuck hepatitis virus (WHV) (Summers et al. 1978), ground squirrel hepatitis virus (GSHV) (Marion et al. 1980), duck hepatitis B virus (DHBV) (Mason et al. 1980), and heron hepatitis B virus (HHBV) (Sprengel et al. 1988).

Evidence for the mechanism of synthesis of the hepadnavirus genome was obtained from the analysis of viral DNA isolated from infected liver, which revealed besides relaxed circular DNA, covalently closed circular DNA (CCC DNA), and linear, single-stranded DNA species of variable length, which were presumably nascent DNAs (Ruiz-Opazo et al. 1982; Mason et al. 1982; Weiser et al. 1983). The examination of the single-stranded, linear, heterogenous DNA forms isolated from livers of DHBV-infected ducks revealed minus-strand DNA and an absence of plus-strand DNA (Mason et al. 1982). This result was incompatible with a mechanism for minus-strand DNA synthesis by polymerization on a plus-strand DNA template and indicated instead that the synthesis of minus strands occurred by an unusual mechanism. In 1982, Summers and Mason demonstrated that minus-strand DNA was synthesized by reverse transcription of an RNA template. A detailed model of this process was finally developed as a result of the precise structural analyses of the RNA template for reverse transcription (Büscher et al. 1985; Enders et al. 1985) and the initiation sites for minus- and plus-strand DNA (Lien et al. 1986, 1987; Seeger et al. 1986; Will et al. 1987). The current status of this model is reviewed below.

The development of tissue culture systems for the propagation of infectious hepadnaviruses greatly facilitated the analysis of the mechanism for viral DNA synthesis. Tuttleman et al. (1986a) described conditions under which primary duck hepatocytes could be infected with DHBV. Similarly, primary human and woodchuck hepatocyte cultures support the replication of HBV and WHV or GSHV, respectively (Theze et al. 1987; Gripon et al. 1988; Rijnitjes et al. 1988; Ochiya et al. 1989; Aldrich et al. 1989). Several laboratories identified hepatoma cell lines that are competent for the expression of HBV and other hepadnaviruses following transfection with cloned viral DNA (Sureau et al. 1987; Tsurimoto et al. 1987; Sells et al. 1987; Sells et al. 1988; Yaginuma et al. 1987; Chang et al. 1987;

HIRSCH et al. 1988; PUGH et al. 1988; SHIH et al. 1989). Although both types of tissue culture systems produce infectious virus, de novo infection appears to be restricted to primary hepatocyte cultures (ACS et al. 1987; SUREAU et al. 1988).

2 Covalently Closed Circular DNA

2.1 Formation of CCC DNA

CCC DNA can be detected in the liver within 6-16 h after infection with DHBV (MASON et al. 1983; TAGAWA et al. 1986). The synthesis of this DNA species precedes the expression of other viral RNA and DNA intermediates. Therefore, it is believed that virion DNA is directly converted into CCC DNA during new rounds of infection. This CCC DNA accumulates in the nuclei of infected hepatocytes where it serves as a template for the transcription of viral RNAs (MILLER and ROBINSON 1984; TAGAWA et al. 1986). The role of CCC DNA as a template for viral RNA synthesis (see chapter by H. Schaller and M. Fischer, this volume) is suggested by the observation that this DNA species and a form of the virion relaxed circular DNA lacking the terminal protein at the 5′ end of minus-strand DNA (WU and MASON, unpublished observation) are apparently the only nuclear species present in chronically infected cells (MILLER and ROBINSON 1984; TAGAWA et al. 1986; PETCU and MASON, unpublished work). Limited digestion of CCC DNA released from liver nuclei prepared from DHBV-infected ducks with microccocal nuclease produces DNA fragments differing in length by regular increments, suggesting that CCC DNA is organized in a nucleoprotein complex (G. SHERMAN, X. HUNG, J. NEWBOLD, personal communication). The significance of this observation for the stability of CCC DNA and for gene expression is not known.

What is the mechanism of the conversion of virion DNA into CCC DNA? The formation of CCC DNA requires(i) the dissociation of the terminal protein (TP) from the 5′ end of minus-strand DNA and the removal of one copy of the terminally redundant segment, termed "r", on minus-strand DNA (Sect. 4.2, Fig. 1A); (ii) the ligation of the 5′ and 3′ ends of minus-strand DNA; and (iii) the removal of the RNA primer from the 5′ end of plus-strand DNA (Sect. 5.1) and the completion of plus-strand DNA synthesis (Fig. 1A, B).

Since minus-strand DNA is the template for the synthesis of pregenomic RNA, the precise ligation of its 5′ and 3′ ends is critical for the maintenance of the viral genome. The formation of a covalently closed minus-strand DNA can be explained by at least two mechanisms, one depending on cellular DNA modifying enzymes, the other on the TP at the 5′ end of minus-strand DNA. The first model requires the removal of the 5′ end of minus-strand DNA including the TP with the help of endonuclease activity and the ligation of the 5′ and 3′ ends. Depending on the position of the deproteinized 5′ end, DNA polymerase activity may be necessary for the extension of the 3′ end to facilitate the ligation of the two ends.

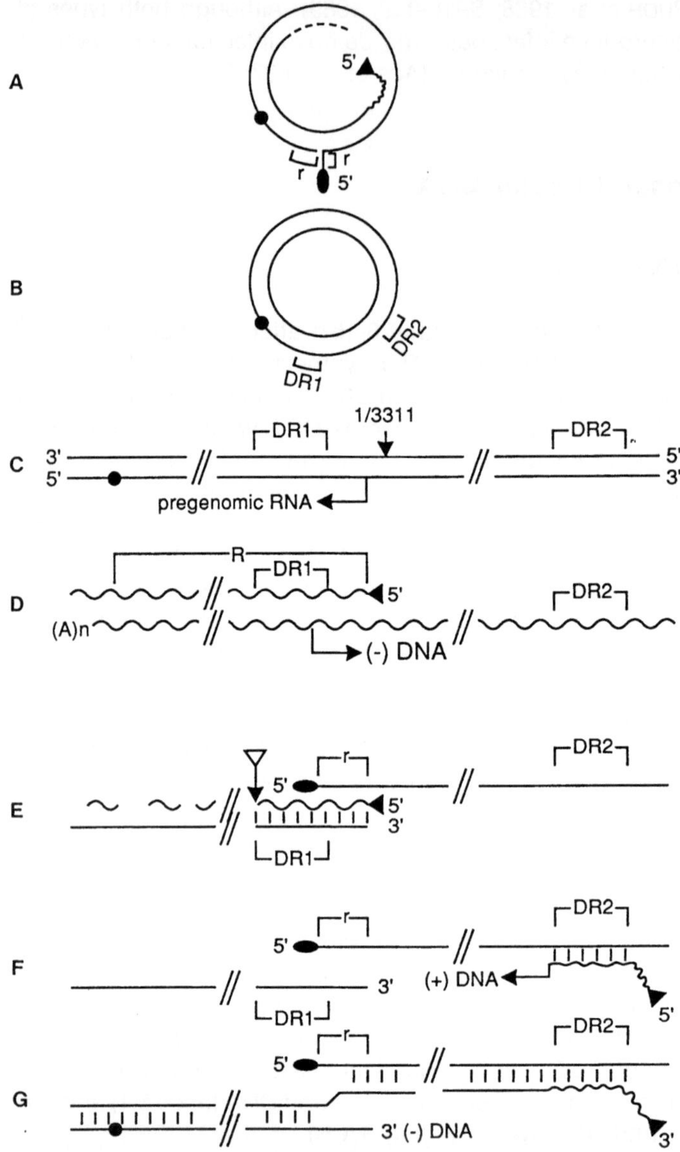

Fig. 1. Model for the synthesis of hepadnaviral DNA. Relaxed circular virion DNA (**A**) is converted into CCC DNA (**B**) during a new round of infection. The ca. 3.3 kb-long CCC DNA contains one copy each of direct repeat 1 (*DR1*) and *DR2* and serves as the template for the synthesis of pregenomic RNA (**C**). Pregenomic RNA (*wavy lines*) initiates ca. 5 nucleotides upstream of DR1 and terminates near a polyadenylation signal (●) downstream of DR1, creating terminally redundant ends, termed "*R*" (**D**) (the cap structure at the 5′ end of the RNA is indicated by the *solid triangle*). Minus-strand DNA synthesis begins at the copy of DR1 close to the 3′ end of pregenomic RNA (**D**) and proceeds to the 5′ end of the RNA template, leading to a ca. 9 nucleotide-long terminal redundancy, termed "*r*" (**E**). As a consequence of this reaction the 5′ end

(*continued*)

The second model requires the TP for the removal of the r region at the 3′ end of minus-strand DNA and the ligation of the 5′ and 3′ ends following hydrolysis of the protein-DNA bond, a reaction carried out by the A protein of bacteriophage ϕX174 for the circularization of the single-stranded DNA genome (ROTH et al. 1984; BROWN et al. 1984). While the first model primarily depends on cellular DNA modifying enzymes and requires DNA synthesis, the second model relies on an enzymatic activity of the TP and does not require DNA synthesis. The observation that foscarnet, an inhibitor of the viral DNA polymerase, does not block the conversion of virion DNA into CCC DNA suggests that the viral DNA polymerase is not required for the completion of minus-strand DNA (MASON et al. 1987a).

2.2 Amplification and Regulation of CCC DNA Synthesis

TUTTLEMAN et al. (1986b) demonstrated that in primary duck hepatocytes infected in vitro with DHBV, CCC DNA was derived not only from infecting viral DNA as described above but also from molecules synthesized de novo early during infection. De novo synthesized CCC DNA was found to be present 3–4 days after infection at levels estimated to be in 50-fold excess of that derived from the infecting virus. Since no production of infectious virus could be demonstrated in the culture supernatants, it was concluded that the amplification of CCC DNA they observed had occurred through an intracellular pathway not involving infection of new cells, Subsequent studies with an inhibitor of viral attachment and uptake, suramin (PETCU et al. 1988), or with a blocking antiserum provided supporting evidence that a second round of infection was not required for de novo synthesis of CCC DNA (WU et al. 1990). Genetic studies revealed that viral envelope proteins were likewise not required for CCC DNA amplification, thereby indicating that virus assembly was not involved. Several lines of evidence, therefore, support the idea that CCC DNA amplification is an intracellular process.

Amplification of CCC DNA appears to be important for subsequent virus production under some conditions. For example, development of high levels of viral replicative intermediates in infected primary hepatocyte cultures can be inhibited by incorporation of 5-bromodeoxyuridine during CCC DNA amplification (TUTTLEMAN et al. 1986b). In addition, enhanced levels of CCC DNA are associated with a more rapid accumulation of replicative viral DNA in primary

◄ of minus-strand DNA becomes covalently, linked to the terminal protein depicted by the *solid oval symbol*. Pregenomic RNA is degraded during minus-strand DNA synthesis by an RNAseH-like activity (**E**). Plus-strand DNA synthesis initiates at the first nucleotide downstream of DR2 (**F**). It is primed by an ca. 18-nucleotide long RNA primer derived from the 5′ end of pregenomic RNA (**E**). The 3′ end of the primer is determined by a RNAseH activity (indicated by an *open triangle*) prior to its transfer to DR2 (**E,F**). Plus-strand DNA synthesis proceeds to the 5′ end of minus-strand DNA, where a template switch to the 3′ end of minus-strand DNA is necessary for the continuation of DNA synthesis (**G**). The template switch is most likely facilitated by the r region at the 5′ and 3′ ends of minus-strand DNA. As a consequence of this reaction, virion DNA assumes a relaxed circular confirmation (**A**) hepatocytes infected with a DHBV genome which fails to suppress amplification (SUMMERS et al. 1990). Whether the effect of amplification is limited simply to supplying more copies of the transcriptional template or whether amplified CCC DNA is transcriptionally different from that derived from infecting viral DNA is not known.

TUTTLEMAN et al. (1986) investigated the mechanism of CCC DNA synthesis during amplification in primary duck hepatocytes congenitally infected with DHBV and maintained in serum-containing medium. Under these culture conditions, levels of CCC DNA were unstable in the infected cells and could increase as much as 100-fold, depending on the cell density, without apparent damage to the cell layer. As in early amplification, this increase was not inhibited by suramin (WU et al. 1990), pointing to the likelihood of an intracellular process similar to that observed at the initiation of infection. Using density-labelling experiments, it was shown that CCC DNA amplification occurred through the reverse transcription pathway, presumably by utilization of DNA-containing nucleocapsids derived from the cytoplasm to augment the nuclear levels of CCC DNA. The same process was thought to occur during early amplification since de novo synthesis of CCC DNA was shown to occur in the absence of any detectable semiconservative replication of CCC DNA derived from the infecting viral DNA (TUTTLEMAN et al. 1986).

Infected cells in culture therefore apparently have the ability to synthesize moderate and sometimes large amounts of CCC DNA in the absence of participation by extracellular virus; yet, in general, the levels of CCC DNA that have been reported to occur in infected liver are relatively low. In a limited survey of ducklings infected with DHBV, under conditions in which every hepatocyte would have been expected to be infected, CCC DNA copy numbers of ca. 3–8 were estimated (PETCU and MASON, unpublished observations). A separate study of a WHV-infected woodchuck revealed a copy number of ca. 50 (J. SUMMERS, unpublished observations). While CCC DNA levels in primary hepatocyte cultures may reach levels of greater than 10^3 per cell in serum-containing medium, culture conditions that more closely maintain the cells in a differentiated state (serum-free medium supplemented with dimethyl sulfoxide) result in the maintenance of low steady state levels of less than 10 copies per infected cell (PUGH and SUMMERS 1989; SUMMERS et al. 1990; J. SUMMERS, unpublished observations). These observations suggest that controls on CCC DNA amplification may limit its levels according to the physiological state of the hepatocyte both in vivo and in vitro.

In a recent study, it was shown that viral envelope protein expression can dramatically suppress CCC DNA amplification in cultured duck hepatocytes maintained in dimethyl sulfoxide-containing medium (SUMMERS et al. 1990). In primary cells infected with a DHBV mutant genome defective in the production of both PreS (P35) and S (P17) envelope proteins, CCC DNA amplification proceeded at an accelerated rate for at least 10 days post infection. In contrast, wild-type genomes showed evidence of suppression of amplification within 4–7 days post infection, and suppression was complete after 7 days. This lack of control exhibited by the mutant genome resulted in the accumulation of 20–30-fold higher levels of CCC DNA than in wild-type infection, normalized to the amount of replicative forms in the infected cultures. Moreover, infection by the mutant virus resulted in only transient expression of DHBV replication in the culture, possibly due to death of the mutant-infected cells. Mutation of the P35 coding region alone has been shown to be sufficient to destroy control of CCC DNA amplification (SUMMERS et al., unpublished observations). The mechanism of suppression of CCC DNA synthesis by P35 is unknown; however, it was found (SUMMERS et al., unpublished observations) that a defect in P35 production also prevents packaging of nucleocapsids into enveloped particles or their secretion from transfected hepatoma cells. This observation suggests that a physical binding of P35 to nucleocapsids may play a role in their assembly into envelopes, thereby preventing their utilization for CCC DNA synthesis in hepatocytes.

It is interesting that the ability of hepadnaviruses to synthesize viral DNA genomes in persistently infected cells enables the infected cells to amplify CCC DNA pools and thereby replenish levels of transcriptional template lost by cell turnover and dilution during cell division. However, amplification is a potentially cytopathic mode of replication in the absence of controls. Apparently, suppression of amplification by envelope proteins is a feedback control for suppressing this cytopathic potential while at the same time maintaining persistent levels of virus production by the cell.

While envelope proteins appear to act at the level of limiting the rate of CCC DNA synthesis, it is not known whether this or other controls can reduce the levels of CC DNA in hepatocytes once high levels are achieved. Reduction of the CCC DNA pool in a nongrowing cell would require degradation or export of the CCC DNA from the nucleus, and this process could potentially lead to partial or complete loss of the ability of an infected cell to transcribe pregenomes and mRNAs and eventually to the loss of virus production. For practical reasons, it will be of great interest to determine whether such down-regulatory controls exist.

3 Pregenomic RNA

3.1 Structure of Pregenomic RNA

CCC DNA is the presumptive template for the transcription of the template for reverse transcription, termed pregenomic RNA (chapter by H. Schaller and M. Fischer, this volume). Although direct evidence is still lacking to confirm this notion, the possibility is further supported by the fact that CCC DNA preparations from cloned viral DNAs are infectious in vivo (WILL et al. 1982; SEEGER et al. 1984; SPRENGEL et al. 1984). Information about the structure of pregenomic RNA is derived from the analysis of viral RNAs isolated from the livers of infected animals, from the characterization of virion DNA, and to some extent from the analysis of RNA isolated from cytoplasmic core particles. Viral RNA isolated from the livers of infected animals contains a ca. 3.5-kb long RNA species and one or two shorter transcripts, ca. 2.1–2.5-kb in length (ENDERS et al. 1985; MOROY et al. 1985; BÜSCHER et al. 1985; chapter by H. SCHALLER and M. FISCHER, this volume). Since replication by reverse transcription requires an RNA template that contains all the information encoded by the viral DNA genome, the 3.5-kb RNA species is the only detectable transcript with the physical structure to function as a replicative intermediate. Since the 3.5-kb transcript has heterogeneous 5′ ends, the nature of pregenomic RNA cannot be inferred solely from the analysis of cytoplasmic RNA. The identity of the 5′ end of pregenomic RNA has been inferred from an analysis of the 5′ ends of plus-strand DNA. These DNA strands are covalently linked to a capped RNA oligomer apparently derived from the 5′ end of one of the major 3.5-kb RNAs (LIEN et al. 1986; SEEGER et al. 1986). As will be discussed below (Sect. 5), since viral DNA synthesis occurs within nucleocapsids, this RNA species appears to be the most frequently packaged template for reverse transcription within the nucleocapsid. Finally, the analogous transcript is found in preparations of cytoplasmic core particles isolated from GSHV-infected livers (ENDERS et al. 1987; LAVINE et al. 1989).

The synthesis of pregenomic RNA leads to a product with terminally redundant ends (R) that exceeds the size of its CCC DNA template by the length of exactly one copy of the terminal repetition (BÜSCHER et al. 1985; ENDERS et al. 1985; chapter by H. SCHALLER and M. FISCHER, this volume; Fig. 1D). An 11–12-nucleotide long sequence termed direct repeat 1 (DR1) is located at both the 5′ end and near the 3′ end of pregenomic RNA. A third copy of the repeat, termed DR2, is located upstream of DR1 close to the 3′ end of pregenomic RNA. As will be discussed below, DR1 and DR2 bear signals important for the initiation of minus- and plus-strand DNA synthesis.

The biological activity of this RNA species to direct viral DNA synthesis has indirectly been validated by experiments in which a cDNA equivalent of pregenomic RNA of HBV and WHV has been fused to heterologous promoters (JUNKER et al. 1987; SEEGER et al. 1989). Transfection of such constructs into cultured cells can lead to the production of infectious virus particles. Most importantly, with these constructs expression of hepadnaviruses is no longer restricted to hepatoma cells. The expression of WHV in fibroblasts leads to the production of infectious virus as determined by the inoculation of susceptible woodchucks (SEEGER et al. 1989). From these experiments it can be inferred that the steps in the hepadnaviral replication cycle that follow the synthesis of pregenomic RNA do not depend on hepatocyte-specific functions.

3.2 Packaging of Pregenomic RNA

To synthesize the viral genome, pregenomic RNA is packaged along with the viral polymerase into core particles. This process most likely requires the interaction of *cis*-acting sequences on pregenomic RNA with components on the core subunits. Alternatively, it is possible that pregenomic RNA and polymerase are encapsidated as a complex where the polymerase bears the signals for packaging (Sect. 6).

ENDERS et al. provided evidence that the 5′ end of GSHV RNA provides signals for encapsidation by demonstrating that the shortest of the three 3.5-kb transcripts, pregenomic RNA, is predominantly packaged into core particles (ENDERS et al. 1987). However, it now appears that the packaging of the longer transcripts is suppressed as a consequence of the translation of the precore gene. HBV and WHV variants with mutations in the initiation codon of the precore gene can package the longer 3.5-kb RNA transcripts (JUNKER et al. 1987; C. SEEGER, unpublished results), suggesting that translational usage of the RNA plays a role in whether it is packaged into nucleocapsids. Recently, M. NIEPMANN-JUNKER et al. reported that a hybrid RNA transcript composed of sequences derived from HBV and the β-galactosidase gene is incorporated into cytoplasmic core particles (NIEPMANN-JUNKER et al. 1990). The HBV fragment in this construct corresponded to the 5′ end of pregenomic RNA including the precore and a portion of the core regions. This result suggests that sequences near the 5′ end of this hybrid RNA, but found at both the 5′ and 3′ ends of the pregenomic RNA, bear the signals for the packaging of the viral RNA. A detailed genetic analysis of these regions on the pregenome should reveal the sequence motif(s) important for its encapsidation.

4 Minus-Strand DNA Synthesis

4.1 Template for Minus-Strand DNA Synthesis

Infected livers contain partially duplexed and free minus-strand DNA species of varying length, ranging from ca. 3-kb to a few hundred bases (MASON et al. 1982; WEISER et al. 1983). A similar pattern is observed when DNA is isolated from viral core particles purified from the cytoplasm of infected hepatocytes (SUMMERS and MASON 1982). Moreover, the incubation of core particles with Mg^{2+} and nucleotide triphosphates leads to the extension of minus- and plus-strand DNA, indicating that core particles contain a DNA polymerase activity similar to the one identified in virions purified from sera and that core particles are the site of viral DNA synthesis (SUMMERS and MASON 1982). At least three observations indicate that RNA is the template for the synthesis of minus-strand DNA. First, the incubation of core particles with actinomycin D, an inhibitor of DNA-directed DNA synthesis, inhibits plus-strand DNA synthesis but does not interfere with the extension of minus strands. Second, the sedimentation analysis of the products obtained from reactions in the presence of actinomycin D reveals that minus-strand DNA is base-paired with RNA. Third, during minus-strand DNA synthesis the RNA portion in RNA:DNA hybrids is progressively reduced, most likely by an RNAseH-like activity known to be associated with the reverse transcriptases of other retroid elements (VARMUS and SWANSTROM 1984; Sect. 7). This notion has recently been supported by the construction of HBV variants with mutations in the polymerase gene, which interfere with the RNAseH activity (RADZIWILL et al. 1990; Sect. 6).

4.2 Structure of Minus-Strand DNA

The 5′ end of minus-strand DNA is covalently attached to a protein, termed TP (GERLICH and ROBINSON 1980; Sect. 6). The DNA-protein bond is stable in 1M NaOH at 95 °C (BARTENSCHLAGER and SCHALLER 1988), conditions under which phosphoseryl and phosphothreonyl linkages are cleaved. Therefore, it has been proposed that the nucleotide-peptide bond between TP and minus-strand DNA is a phosphodiester linkage similar to the phosphotyrosyl bond between the poliovirus VPg protein and viral RNA (WIMMER 1982; ROTHBERG et al. 1978). Since alkaline stability is also a characteristic of certain phosphoamide bonds, this notion awaits further confirmation (SHABAROVA 1970; GUMPORT and LEHMAN 1971).

Based on direct nucleotide sequence analysis of deproteinized virion DNA, the 5′ end of minus-strand DNA of DHBV was mapped to a dG residue at the third nucleotide of DR1 (LIEN et al. 1987). Similar results were obtained from the analysis of minus-strand DNA from GSHV, WHV, and HBV (SEEGER et al. 1986; WILL et al. 1987). These results indicate that the TP is linked to DNA via dGMP. The 3′ end of minus-strand DNA coincides with the 5′ end of the shortest of the three 3.5-kb RNAs. Thus, the 3′ end of the minus strand is specified by "run-off" when the 5′ end of the RNA template is reached during reverse transcription (SEEGER et al. 1986; LIEN et al. 1987; WILL et al. 1987). As a consequence of the position of the 5′ and 3′ ends, minus-strand DNA has a 9–10-nucleotide long terminal redundancy (r) (Fig. 1F). As will be discussed below, the r region is believed to play a role during the template switch required for the synthesis of plus-strand DNA.

4.3 Initiation of Minus-Strand DNA Synthesis

Since the 5′ end of minus-strand DNA is located at DR1 within the terminal redundant region of pregenomic RNA (R, Fig. 1D), the initiation site for minus-strand DNA synthesis on pregenomic RNA has remained ambiguous. Initiation of DNA synthesis at DR1 close to the 5′ end of pregenomic RNA would require a template switch of a nascent minus-strand DNA to the 3′ end of the RNA template, whereas initiation at DR1 near the 3′ end of the RNA template would allow continuous DNA synthesis. To distinguish between the two possibilities, it is necessary to manipulate the two copies of DR1 independently of each other, which requires the resolution of the R region in pregenomic RNA at the DNA level. For this purpose, SEEGER and MARAGOS (1990) fused a cDNA equivalent of pregenomic RNA of WHV with the cytomegalovirus immediate early promoter and introduced short insertions near both copies of DR1. The analysis of virion DNA expressed with these constructs showed that the 5′ ends of minus-strand DNA map to the DR1 sequence at the 3′ end of pregenomic RNA (Fig. 1D; SEEGER and MARAGOS 1990). Hence, the synthesis of minus-strand DNA from the pregenomic RNA template can proceed continuously from the DR1 to the 5′ end of the pregenome. Thus, unlike in retroviruses (TAYLOR and ILLMENSEE 1975; VARMUS and SWANSTROM 1985) and caulimoviruses (COVEY et al. 1983; MASON et al. 1987b), the strategy employed by hepadnaviruses for minus-strand DNA synthesis does not require the transfer on nascent minus-strand DNA from the 5′ end of the RNA template to its 3′ end (Fig. 1D, E).

The mechanism by which minus-strand DNA synthesis is primed is not yet understood. Initial experiments aimed at an identification of the nucleotide sequences required in *cis* for minus-strand DNA synthesis revealed a ca. 20-nucleotide long sequence motif flanking DR1 at the 3′ end of pregenomic RNA that is essential for the synthesis of the 3.3-kb long DNA species (SEEGER and MARAGOS 1990). This notion is supported by the observation that deletions of nucleotide sequences encoding the signal abrogate the synthesis of 3.3-kb minus-strand DNA. Insertion of a synthetic signal at the 3′ end of pregenomic RNA directs the synthesis of minus-strand DNAs, which are covalently attached to protein. However, based on deletion analyses of the WHV genome it appears that additional nucleotide sequences are required for minus-strand DNA synthesis. They are located between the termination site for the translation of the viral envelope proteins and DR1 (J. PLATKO, J. MARAGOS, and C. SEEGER, unpublished observation). The role of these signal sequences for minus-strand DNA synthesis remains to be elucidated.

The detection of protein covalently attached to the 5′ end of minus-strand DNA of HBV and all other known hepadnaviruses led to the speculation that viral DNA synthesis is primed by protein rather than by RNA (GERLICH and ROBINSON 1980). The evidence favoring this assumption is primarily based on the observation that short, nascent minus-strand DNA species isolated from cytoplasmic core particles in DHBV-infected duck livers are covalently linked to the TP (MOLNAR-KIMBER et al. 1983). The presence of protein covalently linked to minus-strand DNA is certainly indicative of a protein priming mechanism, e.g., as exemplified by the adenovirus or bacteriophage ϕ29 systems (STILLMANN 1989; SALAS 1983). However, it cannot be excluded that protein is added to nascent minus-strand DNA following the initiation of DNA synthesis. For instance, purified poliovirus VPg protein can attach in vitro to an internal site of a viral RNA template by a cleavage-ligation reaction (TOBIN et al. 1989).

The results described above indicate that the signals for initiation of minus-strand DNA synthesis and for the addition of protein to the 5′ end of minus-strand DNA are closely linked and may not be genetically separable Since viral sequences on pregenomic RNA between DR1 and the polyadenylation site are not required for minus-strand DNA synthesis, a model in which priming of minus-strand DNA occurs in the 3′ direction of DR1, e.g., by an RNA primer, can be revoked (SEEGER and MARAGOS 1990). A more definitive explanation for the mechanism of minus-strand DNA synthesis requires an in vitro system in which the priming reaction can be reconstituted with purified viral components.

5 Plus-Strand DNA

5.1 Structure of Plus-Strand DNA

The first structural analyses of hepadnavirus genomes revealed that the 5′ ends of minus- and plus-strand DNAs of HBV are separated by ca. 200 nucleotides (SUMMERS et al. 1975). The failure of both 5′ ends to accept phosphate after incubation with phosphatase to remove any terminal phosphates and polynucleotide kinase to catalyze, the addition of radioactive phosphate suggested that like the minus strand, the 5′ ends of plus-strand DNAs are also modified (GERLICH and ROBINSON 1980). This notion has been confirmed by the studies of LIEN et al. who identified a 17–18-nucleotide long RNA oligomer covalently attached to plus-strand DNAs of DHBV (LIEN et al. 1986). The authors also demonstrated that the 5′ ends of the RNA molecules are modified with a cap structure, common to most eukaryotic mRNAs (LIEN et al. 1986; Fig. 1F). A cap structure would not be susceptible to phosphorylation by polynucleotide kinase.

The 5′ ends of RNAse-treated, plus-strand DNAs of avian and mammalian hepadnaviruses map to the first nucleotide downstream of DR2 (LIEN et al. 1986; SEEGER et al. 1986). Plus-strand DNA synthesis proceeds to the 5′ end of minus-strand DNA. As a consequence of the position of their 5′ ends, nascent plus strands must switch the DNA template as they reach the 5′ end of minus-strand DNA. This step requires the dissociation of the 3′ ends from the 5′ terminal portion of minus-strand DNA and their association with the 3′ ends of minus strands. The transfer is believed to be facilitated by the short terminal redundancy (r) on the minus-strand DNA (Fig. 1E, F). The protein at the 5′ end of minus-strand DNA may provide a scaffold for this event, e.g., by keeping the two ends of minus-strand DNA in close proximity. However, the determinants of this reaction await experimental characterization. The continuation of plus-strand DNA synthesis beyond the 5′ end of its template accounts for the formation of the cohesive overlap structure of virion DNA (Fig. 1A, G). For reasons which are not yet understood, DNA synthesis in the mammalian hepadnaviruses terminates 30%–50% short of its completion whereas in DHBV, plus-strand DNA synthesis appears to be almost complete, terminating at or a few nucleotides in the 5′ direction of DR2 (ROBINSON et al. 1974; SUMMERS et al. 1975; GANEM et al. 1982; LIEN et al. 1987).

5.2 Priming of Plus-Strand DNA Synthesis

As a consequence of the reverse transcription reaction, minus-strand DNA is the template for plus-strand DNA synthesis, and therefore, the priming reactions for the synthesis of the two DNA strands are temporally separated from each other and occur by different mechanisms. In analogy to the retroviral model, in which plus-strand DNA synthesis is primed by an RNA oligomer derived from the viral RNA genome, it has been proposed that plus-strand DNA synthesis of hepadnaviruses is primed by the RNA oligomer found at the 5′ end of plus-strand DNA (LIEN et al. 1986; SEEGER et al. 1986; WILL et al. 1987). In retroviruses, the RNA primer is determined by the reverse transcriptase-associated RNAseH activity and originates from an RNA segment abutting the initiation site for plus-strand DNA synthesis. In contrast, the RNA oligomer attached to plus strands of hepadnaviruses originates from the 5′ end of pregenomic RNA and bears a copy of DR1 (LIEN et al. 1986; SEEGER et al. 1986; Fig. 1E, F). Therefore, it has been proposed that following the completion of minus-strand DNA synthesis, the RNA primer is transported to DR2 on minus-strand DNA to prime the synthesis of plus-strand DNA. This model has been confirmed with the analysis of virion DNA expressed by variants of the GSHV and WHV genomes. First, a mutation introduced into DR1 leads to the expression of viral genomes which carry the same mutation in the RNA primer attached to plus-strand DNA (SEEGER et al. 1986). This finding proves that the synthesis of the DR portion of the RNA primer is directed by DR1 and not by DR2. Second, nucleotide insertions or deletions into viral genomes that alter the position of the 5′ ends of pregenomic RNA lead to the expression of progeny with the corresponding changes at the 5′ end of the RNA primer. Hence, the position of the 5′ end of pregenomic RNA dictates the position of the 5′ end of the RNA primer attached to plus-strand DNA (SEEGER and MARAGOS 1989). Third, mutations in DR1 that destroy the homology between the direct repeats abrogate plus-strand DNA synthesis. Such mutants can be rescued by introducing the corresponding nucleotide changes into DR2, indicating that the annealing of the RNA primer is an important prerequisite for the priming reaction (A. GLASER and C. SEEGER, unpublished results).

As a consequence of the mechanism for plus-strand DNA synthesis, the 5′ portion of pregenomic RNA has a dual function: First it serves as a template for the synthesis of the 3′ end of minus-strand DNA and then as a primer for the synthesis of plus-strand DNA. This model predicts that initiation of plus-strand DNA synthesis at DR2 can only occur subsequent to the completion of minus-strand DNA synthesis. Alternatively, two RNA templates would be required for the synthesis of one copy of the viral genome.

The mechanism for the priming of plus strands requires several biochemical steps, which include (a) the formation of the exact 3′ end of the RNA primer, (b) the dissociation of the primer from the 3′ end of minus-strand DNA, and (c) the transport and annealing of the RNA oligomer to DR2 on minus-strand DNA.

The 3′ end of the RNA primer is most likely determined by RNAseH activity prior to its transfer to DR2 (SEEGER and MARAGOS 1989; Fig. 1E). The signals on pregenomic RNA that determine the cleavage site must reside downstream of it. This notion is based on the observation that changes in the nucleotide sequence composition of the DR1 of WHV do not interfere with primer formation (A.L. GLASER and C. SEEGER, unpublished results). Similarly, base changes introduced at the RNA cleavage site do not affect the formation of the RNA primer, and pregenomic RNAs with small base deletions at their 5′ ends yield RNA primers with normal 3′ ends (SEEGER and MARAGOS 1989). In contrast, mutations in the DHBV genome immediately 3′ of DR1 can abolish priming of plus-strand DNA synthesis at DR2 and lead to the synthesis of double-stranded linear genomes (S. STRAPANS and D. GANEM, personal communication). It is not yet known whether these lesions affect the cleavage or the translocation of the RNA primer to DR2. At least two mechanisms can be proposed that explain the transfer of the RNA primer to the primer binding site at DR2. The first implicates the viral polymerase as a vehicle for the translocation of the RNA oligomer; the second depends on the physical proximity of the repeats on minus-strand DNA. Such an arrangement may promote an intramolecular transfer of the RNA primer from DR1 to DR2.

6 The Viral DNA Polymerase

Evidence that the viral *pol* gene product encodes an RNA-directed DNA polymerase activity has been inferred from amino acid homologies discovered between *pol* gene products of hepadnaviruses and retroviruses (TOH et al. 1983). The biochemical analysis of the hepadnavirus DNA polymerase is mainly hampered by the failure to solubilize this enzyme from virions. The deficiency of the HBV polymerase to accept exogenous templates can be extended to the DHBV system, in which the reverse transcriptase fails to switch to non viral templates even after the permeabilization of viral core particles by treatment incubation in solutions with low pH (KAPLAN et al. 1973; RADZIWILL et al. 1988). Although enzymatically active reverse transcriptase can be produced in *Escherichia coli* by the expression of retroviral *pol* genes, attempts to express the hepadnavirus enzyme in its active form have so far been without success (TANESE et al. 1985, 1986; KOTEWICZ et al. 1985).

The possibility of expressing hepadnaviruses in tissue culture cells transfected with cloned viral DNA now provides the major system for a functional analysis of the polymerase gene products. The most important finding is that the polymerase gene of DHBV is expressed independently from the viral core gene product and can function in *trans* to synthesize viral DNA. CHANG et al. and SCHLICHT et al. demonstrated that the DHBV *pol* gene is translated from an internal initiation (AUG) codon near the 5′ end of the open reading frame (SCHLICHT et al. 1989; CHANG et al. 1989). Mutations that alter this initiation codon abrogate the production of functional *pol* gene products. The initiation site conserved among the sequenced viral genomes, and the results obtained from the DHBV system can be extended to the other hepadnaviruses (ROYCHOUDHURY and SHIH 1990).

Evidence has been provided that the polymerase gene encodes the terminal protein at the 5′ end of minus-strand DNA (BARTENSCHLAGER and SCHALLER 1988). This conclusion is based on experiments demonstrating that DHBV DNA can be immunoprecipitated with antibodies raised against different domains of the *pol* gene product from preparations containing cytoplasmic core particles. After partial proteolytic degradation of the replication complex, antibodies raised against the amino-terminal domain of the *pol* gene product can still immunoprecipitate viral DNA. This result suggests that the peptide linked to DNA originates from the amino-terminal portion of the viral polymerase protein. More extensive characterization of the protein component attached to DNA will be necessary to validate this conclusion further.

The carboxyl-terminal domain of the polymerase protein may encode the reverse transcriptase-associated RNAseH activity (RADZIWILL et al. 1990). This region shows some homology with the *E. coli* and Moloney murine leukemia virus RNAseH domains. HBV genomes with amino acid substitutions in this region of the *pol* gene are either defective for DNA synthesis or produce only minus-strand DNA, albeit at reduced levels as compared with wild-type DNA. It is possible to conclude from this observation that plus-strand DNA synthesis is blocked due to a deficiency of the RNAseH activity required for the removal of RNA from minus-strand DNA (RADZIWILL et al. 1990).

Despite the recent progress made in understanding polymerase function, several issues remain to be resolved. For example, we do not know the physical structure of the enzymatically active polymerase protein. Is this protein proteolytically cleaved or otherwise modified? The failure to obtain enzymatically active polymerase products in *E. coli* may be a consequence of the requirement for posttranslational modifications that are not catalyzed in bacteria. Moreover, it is not known whether the reverse transcriptase is enzymatically active after completion of translation or whether enzymatic activity depends on other viral components. Since the polymerase is translated independently from the viral capsid protein, a mechanism that is so far without precedent among retroid elements, it must contain signals for its incorporation into the core tegument. Pregenomic RNA or core subunits are the logical candidates for this interaction.

7 Perspectives

With the help of animal models and recently with the development of tissue culture systems, we have obtained a very detailed picture of the mechanism of the synthesis of the hepadnavirus genome. The structures of the major replicative DNA and RNA intermediates have been determined at the nucleotide level, and detailed models for virus replication have been proposed. This work set the stage for a functional analysis of the signals on the viral genome that orchestrate the synthesis of viral progeny. Among them are the signals for the synthesis of CCC DNA, for the packaging of pregenomic RNA, and for the initiation of reverse transcription of minus-strand DNA.

The investigation of the mechanism that governs the maintenance of CCC DNA levels may provide insight into the process which controls the course of viral infection and lead to the identification of the cellular components that play a role in directing viral DNA either to the cytoplasmic membrane or to the nucleus. The assembly of the DNA replication complex requires signals not only on pregenomic RNA for encapsidation but also on the viral reverse transcriptase. What is the sequence of events that leads to the incorporation of the two components into the core tegument, and when does reverse transcription of minus-strand DNA begin? The mechanism for plus-strand DNA synthesis is still enigmatic. The translocation of an RNA primer for DNA synthesis is so far without precedent. Why do hepadnaviruses employ this seemingly complicated mechanism for the priming of plus-strand DNA in view of the much simpler concept adopted by all other known retroid elements? The elucidation of the mechanism for plus-strand DNA synthesis will be difficult unless enzymatically active reverse transcriptase can be produced to carry out the priming reaction in vitro with synthetic RNA and DNA templates. Finally, with the availability of tissue culture systems for the expression of functional reverse transcriptase and antibodies raised against different domains of the *pol* protein, it should be possible to characterize the physical structure of the polymerase products in the cytoplasm and in core particles of virus-producing cells.

Explanations for these and other unresolved problems of the viral replication cycle may provide novel targets for antiviral therapies that are urgently needed for the millions of chronically infected individuals. Although the mchanism for the synthesis of viral DNA follows the general concept used by retroviruses and other retroid elements, DNA synthesis in hepadnaviruses passes through several steps which are unique. Not surprisingly, they pose the major problems in completing our understanding of viral replication.

Acknowledgements. C.S. is supported by Public Health Service grant AI-24972 from the National Institutes of Health. W.S.M. is supported by Public Health Service grants from the National Institutes of Health (AI-18641, CA-06927, and RR-05539) and an appropriation from the Commonwealth of Pennsylvania. J.S. is supported by Public Health Service grant CA-42542 from the National Institutes of Health.

References

Acs G, Sells MA, Purcell RH, Price P, Engle R, Shapiro M, Popper H (1987) Hepatitis B virus produced by transfected Hep G2 cells causes hepatitis in chimpanzees. Proc Natl Acad Sci USA 84: 4641–4644

Aldrich CE, Coates L, Wu T-T, Newbold J, Tennant BC, Summers J, Seeger C, Mason WS (1989) *In vitro* infection of woodchuck hepatocytes with woodchuck hepatitis virus and ground squirrel hepatitis virus. Virology 172: 247–252

Bartenschlager R, Schaller H (1988) The amino-terminal domain of the hepadnaviral P-gene encodes the terminal protein (genome-linked protein) believed to prime reverse transcription. EMBO J 7: 4185–4192

Brown DR, Roth MJ, Reinberg D, Hurwitz J (1984) Analysis of bacteriphage phjX174 gene A protein-mediated termination and reinitiation of phiX DNA synthesis. J Biol Chem 259: 10545–10555

Büscher M, Reiser W, Will H, Schaller H (1985) Transcripts and the putative RNA pregenome of duck hepatitis B virus implications for reverse transcription. Cell 40: 717–724

Chang CM, Jeng KS, Hu CP, Lo SJ, Su TS, Ting LP, Chou CK, Ha SH, Pfaff E, Salfeld J (1987) Production of hepatitis B virus *in vitro* by transient expression of cloned HBV DNA in a hepatoma cell line. EMBO J 6: 675–680

Chang L-J, Pryciak P, Ganem D, Varmus HE (1989) Biosynthesis of the reverse transcriptase of hepatitis B viruses involves *de novo* translation initiation not ribosomal frameshifting. Nature 337: 364–368

Covey SN, Turner D, Mulder G (1983) A small DNA molecule containing covalently-linked ribonucleotides originates from the large intergenic region of the cauliflower mosaic virus genome. Nucleic Acids Res 11: 251–264

Enders GH, Ganem D, Varmus HE (1985) Mapping the major transcripts of ground squirrel hepatitis virus: The presumptive template for reverse transcriptase is terminally redundant. Cell 42: 297–308

Enders GH, Ganem D, Varmus HE (1987) 5'-Terminal sequences influence the segregation of ground squirrel hepatitis virus RNA into polyribosomes and viral core particles. J Virol 61: 35–41

Ganem D, Greenbaum L, Varmus HE (1982) Virion DNA of ground squirrel hepatitis virus structural charactrization and molecular cloning. J Virol 44: 374–383

Gerlich WH, Robinson WS (1980) Hepatitis B virus contains protein attached to the 5' terminus of its complete DNA strand. Cell 21: 801–809

Gripon P, Diot C, Theze N, Fourel I, Loreal O, Brechot C, Guguen-Guillouzo C (1988) Hepatitis B virus infection of adult human hepatocytes cultured in the presence of dimethyl sulfoxide. J virol 62: 4136–4143

Gumport RI, Lehman IR (1971) Structure of the DNA ligase-adenylate intermediate Lysine (e-amino)-linked adenosine monophosphoramidate. Proc Natl Acad Sci USA 68: 2559–2563

Hirsch R, Colgrove R, Ganem D (1988) Replication of duck hepatitis B virus in two differentiated human hepatoma cell lines after transfection with cloned viral DNA. Virology 167: 136–142

Junker M, Galle P, Schaller H (1987) Expression and replication of the hepatitis B virus genome under foreign promoter control. Nucleic Acids Res 15: 10117–10132

Kaplan PM, Greenman RL, Gerin JL, Purcell RH, Robinson WS (1973) DNA polymerase associated with human hepatitis B antigen. J Virol 12: 995–1005

Kotewicz ML, D'Alessio JM, Driftmier KM, Blodgett KP, Gerard GF (1985) Cloning and overexpression of Moloney murine leukemia virus reverse transcriptase in *Escherichia coli*. Gene 35: 249–258

Lavine J, Hirsch R, Ganem D (1989) A system for studying the selective encapsidation of hepadnavirus RNA. J Virol 63: 4257–4263

Lien J, Aldrich CE, Mason WS (1986) Evidence that a capped oligoribonucleotide is the primer for duck hepatitis B virus plus strand DNA synthesis. J Virol 57: 229–236

Lien J, Petcu DJ, Aldrich CE, Mason WS (1987) Initiation and termination of duck hepatitis B virus DNA synthesis during virus maturation. J Virol 61: 3832–3840

Marion PL, Oshiro LS, Regnery DC, Scullard GH, Robinson WS (1980) A virus in Beechey ground squirrels that is related to hepatitis B virus of humans. Proc Natl Acad Sci USA 77: 2941–2945

Mason WS, Seal G, Summers J (1980) Virus of Pekin ducks with structural and biological relatedness to human hepatitis B virus. J Virol 36: 829–836

Mason WS, Aldrich C, Summers J, Taylor JM (1982) Asymmetric replication of duck hepatitis B virus DNA in liver cells: Free minus-strand DNA. Proc Natl Acad Sci USA 79: 3997–4001

Mason WS, Halpern MS, England JM, Seal G, Egan J, Coates L, Aldrich C, Summers J (1983) Experimental transmission of duck hepatitis B virus. Virology 131: 375–384

Mason WS, Lien J, Petcu DJ, Coates L, London WT, O'Connell A, Aldrich C, Custer PR (1987a) *In vivo* and *in vitro* studies on duck hepatitis B virus. Replication. In: Robinson WS, Kolke K, Will H (eds) Hepadna viruses. Liss, New York, pp 3–16

Mason WS, Taylor JM, Hull R (1978b) Retroid virus genome replication. Adv virus Res 32: 35–96

Miller RH, Robinson WS (1984) Hepatitis B virus DNA in nuclear and cytoplasmic fractions of infected human liver. Virology 137: 390–399

Molnar-Kimber KL, Summers JW, Taylor JM, Mason WS (1983) Protein covalently bound to minus strand DNA intermediates of duck hepatitis B virus. J Virol 45: 165–172
Molnar-Kimber KL, Summers JW, Mason WS (1984) Mapping of the cohesive overlap of duck hepatitis B virus DNA and of the site of initiation of reverse transcription. J Virol 51: 181–191
Moroy T, Etiemble J, Trepo C, Tiollais P, Buendia MA (1985) Transcription of woodchuck hepatitis virus in the chronically infected liver. EMBO J 4: 1507–1514
Niepmann-Junker M, Bartenschlager R, Schaller H (1990) A short cis-acting sequence is required for hepatitis B virus pregenome encapsidation and sufficient for packaging of foreign RNA. EMBO 9: 3389–3396
Ochiya T, Tsurimoto T, Ueda K, Okubo K, Shiozawa M, Matsubara K (1989) An *in vitro* system for infection with hepatitis B virus that uses primary human fetal hepatocytes. Proc Natl Acad Sci USA 86: 1875–1879
Pasek M, Goto T, Gilbert W, Zink B, Schaller H, MacKay P, Leadbetter G, Murray K (1979) Hepatitis B virus genes and their expression in *E coli*. Nature 282: 575–579
Petcu DJ, Aldrich CE, Coates L, Taylor JM, Mason WS (1988) Suramin inhibits *in vitro* infection by duck hepatitis B virus, Rous sarcoma virus, and hepatitis delta virus. J Virol 167: 385–392
Pugh JC, Yaginuma K, Koike K, Summers J (1988) Duck hepatitis B virus (DHBV) particles produced by transient expression of DHBV DNA in a human hepatoma cell line are infectious *in vitro*. J Virol 62: 3513–3616
Pugh JC, Summers J (1989) Infection and uptake of duck hepatitis B virus by duck hepatocytes maintained in the presence of dimethyl sulfate. J Virol 172: 564–572
Radziwill G, Zentgraf H, Schaller H, Bosch V (1988) The duck hepatitis B virus polymerase is tightly associated with the viral core structure and unable to switch to an exogenous template. Virology 163: 123–132
Radziwill G, Tucker W, Schaller H (1990) Mutational analysis of the hepatitis B virus P gene product domain structure and RNAseH activity. J Virol 64: 613–620
Rijnitjes PJM, Morshage HJ, Yap SH (1988) *In vitro* infection of primary cultures of cryopreserved adult human hepatocytes with hepatitis B virus. Virus Res 10: 95–110
Robinson WS, Clayton DA, Greenman RL (1974) DNA of a human hepatitis B virus candidate. J Virol 14: 384–391
Roth MJ, Brown DR, Hurwitz J (1984) Analysis of bacteriophage phiX174 gene A protein-mediated termination and reinitiaion of phiX DNA synthesis. J Biol Chem 259: 10556–10568
Rothberg PG, Harris TJ, Nomoto A, Wimmer E (1978) The genome-linked protein of picornaviruses, V. O^4-(5'-uridylyl) tyrosine is the bond between the genome-linked protein and the RNA of poliovirus. Proc Natl Acad Sci USA 75: 4868–4872
Roychoudhury S, Shih C (1990) cis Rescue of a mutated reverse transcriptase of human hepatitis B virus by creation of an internal AUG. J Virol 64: 1063–1069
Ruiz-Opazo N, Chakraborty PR, Shafritz DA (1982) Evidence for supercoiled hepatitis B virus DNA in chimpanzee liver and serum Dane particles: possible implications in persistent HBV infection. Cell 29: 129–138
Salas M (1983) A new mechanism for the initiation of replication of ϕ29 and adenovirus DNA: priming by the terminal protein. In: Doerfler W (ed) The molecular biology of adenoviruses 1. 30 years of adenovirus research 1953–1983. Springer, Berlin Heidelberg New York, pp 89–106 (Current topics in microbiology and immunology, vol 109)
Sattler F, Robinson WS (1979) Hepatitis B viral DNA molecules have cohesive ends. J Virol 32: 226–233
Schlicht HJ, Radziwill G, Schaller H (1989) Synthesis and encapsidation of duck hepatitis B virus reverse transcriptase do not require formation of core-polymerase fusion proteins. Cell 56: 85–92
Seeger C, Maragos J (1989) Molecular analysis of the function of direct repeats and a polypurine tract for plus-strand DNA priming in woodchuck hepatitis virus. J Virol 63: 1907–1915
Seeger C, Maragos J (1990) Identification and characterization of the woodchuck hepatitis virus origin of DNA replication. J Virol 64: 16–23
Seeger C, Ganem D, Varmus HE (1984) The cloned genome of ground squirrel hepatitis virus is infectious in the animal. Proc Natl Acad Sci USA 81: 5849–5852
Seeger C, Ganem D, Varmus HE (1986) Biochemical and genetic evidence for the hepatitis B virus replication strategy. Science 232: 477–484
Seeger C, Baldwin B, Tennant BC (1989) Expression of infectious woodchukc hepatitis virus in murine and avian fibroblasts. J Virol 63: 4665–4669

Sells MA, Chen M, Acs G (1987) Production of hepatitis B virus particles in hepG2 cells transfected with cloned hepatitis B virus DNA. Proc Natl Acad Sci USA 84: 1005–1009
Sells MA, Zelent AZ, Shvartsman M, Acs G (1988) Replicative intermediates of hepatitis B virus in HepG2 cells that produce infectious virions. J Virol 62: 2836–2844
Shabarova ZA (1970) Synthetic nucleotide-peptides. In: Davidson JN, Cohn W (eds) Progress in nucleic acid research and molecular biology, vol. X. Academic, New York, pp 145–182
Shih C, Li L-S, Roychoudhury S, Ho M-H (1989) *In vitro* propagation of human hepatitis B virus in a rat hepatoma cell line. Proc Natl Acad Sci USA 86: 6323–6327
Sprengel R, Kuhn C, Manso K, Will H (1984) Cloned duck hepatitis B virus DNA is infectious in pekin ducks. J Virol 52: 932–937
Sprengel R, Kaleta EF, Will H (1988) Isolation and characterization of a hepatitis B virus endemic in herons. J Virol 62: 3832–3839
Stillmann B (1989) Initiation of eukaryotic DNA replication *in vitro*. Annu Rev Cell Biol 5: 197–246
Summers J, Mason WS (1982) Replication of the genome of a hepatitis B-like virus by reverse transcription of an RNA intermediate. Cell 29: 403–415
Summers J, O'Connell A, Millman I (1975) Genome of hepatitis B virus restriction enzyme cleavage and structure of DNA extracted from Dane particles. Proc Natl Acad Sci USA 72: 4597–4601
Summers J, Smolec JM, Snyder R (1978) A virus similar to human hepatitis B virus associated with hepatitis and hepatoma in woodchucks. Proc Natl Acad Sci USA 75: 4533–4537
Summers J, Smith PM, Horwich AL (1990) Hepadnaviral envelope proteins regulate covalently closed circular DNA amplification. J Virol 64: 2819–2824
Sureau C, Romet-Lemonne J, Mullins JI, Essex M (1987) Production of hepatitis B virus by a differentiated human hepatoma cell line after transfection with cloned circular HBV DNA. Cell 47: 37–47
Sureau C, Eichberg JW, Hubbard GB, Romet-Lemonne JL, Essex M (1988) A molecularly cloned hepatitis B virus produced *in vitro* is infectious in a chimpanzee. J Virol 62: 3064–3067
Tagawa M, Omata M, Okuda K (1986) Appearance of viral RNA transcripts in the early stage of duck hepatitis B virus infection. Virology 152: 477–482
Takeda N, Kuhn RJ, Yang C, Takegami T, Wimmer E (1986) Initiation of poliovirus plus-strand RNA synthesis in a membrane complex of infected HeLa cells. J Virol 60: 43–53
Tanese N, Roth M, Goff SP (1985) Expression of enzymatically active reverse transcriptase in *Echerichia coli*. Proc Natl Acad Sci USA 82: 4944–4948
Tanese N, Sodrowski J, Haseltine WA, Goff SP (1986) Expression of reverse transcriptase activity of human T-lymphotropic virus type III (HTLV-III/LAV) in *Escherichia coli*. J Virol 59: 743–745
Taylor JM, Illmensee R (1975) Site on the RNA of an avian sarcoma virus at which primer is bound. J Virol 16: 533–558
Theze N, Gripon P, Fourel I, Hantz O, Trepo C, Guguen-Guillouzo C (1987) Maintenance of woodchuck hepatitis virus activity in primary cultures of woodchuck hepatocytes. J Gen Virol 68: 1029–1039
Tobin GJ, Young DC, Flanegan JB (1989) Self-catalyzed linkage of poliovirus terminal protein VPg to poliovirus RNA. Cell 59: 511–519
Toh H, Hayashida H, Miyata T (1983) Sequence homology between retroviral reverse transcriptase and putative polymerases of hepatitis B virus and cauliflower mosaic virus. Nature 305: 827–829
Tsurimoto T, Fujiyama A, Matsubara K (1987) Stable expression and replication of hepatitis B virus genome in an integrated state in a human hepatoma cell line transfected with the cloned viral DNA. Proc Natl Acad Sci USA 84: 444–448
Tuttleman JS, Pugh JC, Summers JW (1986a) *In vitro* experimental infection of primary duck hepatocyte cultures with duck hepatitis B virus. J Virol 58: 17–25
Tuttleman JS, Pourcel C, Summers JW (1986b) Formation of the pool of covalently closed circular viral DNA in hepadnavirus-infected cells. Cell 47: 451–46
Varmus HE, Swanstrom R (1984) Replication of retroviruses. In: Weiss R, Teich N, Varmus HE, Coffin J (eds) RNA tumor viruses. Cold Spring Harbor Laboratory, Cold Spring Harbor, pp 369–512
Varmus HE, Swanstrom R (1985) Replication of retroviruses. In: Weiss R, Teich N, Varmus HE, Coffin J (eds) RNA tumor viruses. Cold Spring Harbor Laboratory, Cold Spring Harbor, pp 75–134
Weiser B, Ganem D, Seeger C, Varmus HE (1983) Closed circular viral DNA and asymmetric heterogeneous forms in livers from animals infected with ground squirrel hepatitis virus. J Virol 48: 1–9
Will H, Cattaneo R, Koch H, Darai G, Schaller H, Schellekens H, van Eer P, Deinhardt F (1982) Cloned HBV DNA causes hepatitis in chimpanzees. Nature 299: 740–742

Will H, Reiser W, Weimer T, Pfaff E, Büscher M, Sprengel R, Cattaneo R, Schaller H (1987) Replication strategy of human hepatitis B virus. J Virol 61: 904–911
Wimmer E (1982) Genome-linked proteins of viruses. Cell 28: 199–201
Wu T-T, Coates L, Aldrich CE, Summers J, Mason WS (1990) In hepatocytes infected with duck hepatitis B virus, the template for viral RNA synthesis is amplified by an intracellular pathway. Virology 175: 255–261
Yaginuma K, Shirakata Y, Kobayashi M, Koike K (1987) Hepatitis B virus (HBV) particles are produced in a cell culture system by transient expression of transfected HBV DNA. Proc Natl Acad Sci USA 84: 2678–2682

Assembly of Hepadnaviral Virions and Subviral Particles

D. Ganem

One of the most distinctive features of hepadnaviral infection is the production by infected cells of multiple types of virus-related particles, including large numbers of noninfections subviral particles as well as complete virions. Although this unusual property was among the first aspects of the viral life cycle to be discovered, it remains one of its least well understood aspects. Why do infected cells produce subviral particles? What factors determine whether virions or subviral particles are produced, and how are their relative levels controlled? What are the cellular pathways involved in the production of each particle type? At present our answers to all of these questions are fragmentary. However, the recent application of biochemical and genetic approaches to the study of hepadnaviral morphogenesis has reawakened interest in this process and helped to frame the overall dimensions of the problem. In the following pages I review what is known about hepadnaviral structure and assembly, with emphasis on recent mutational approaches that examine the role of individual viral gene products in the assembly process. Since the work is most advanced for the human hepatitis B virus (HBV), this review will concentrate on HBV; mention will be made of the animal hepadnaviruses only where significant differences exist or where the corresponding information for HBV is not available.

Departments of Microbiology and Immunology and Medicine, University of California Medical Center, San Francisco, CA 94143-0503, USA

Current Topics in Microbiology and Immunology, Vol. 168

1 Hepadnavirus Structure and Replication: An Overview

Shortly after the discovery of the so-called Australia antigen in the serum of multiply transfused individuals (BLUMBERG et al. 1965), it was found that the immunoreactive material resided in a 20-nm virus-like particle (BAYER et al. 1968; GERIN et al. 1969). For a brief period following its discovery, this particle was mistakenly assumed to be the infectious virion of HBV itself. Soon thereafter, however, electron-microscopic examinations of partially purified, serum-derived preparations of HBV revealed the presence of three morphologically distinct particle types: spheres and filaments of 20-nm diameter and less abundant 42-nm double-shelled particles now known as "Dane particles" (DANE et al. 1970). Immunologic studies demonstrated that all three particle types share on their surface the Australia antigen (now called hepatitis B surface antigen, HBsAg) and hence are clearly of viral origin.

The Dane particle is now known to be the infectious virion of HBV (ROBINSON and LUTTWICK 1976). It consists of an outer shell or envelope comprised of HBsAg, presumably in association with host-derived lipid. Internal to this is the viral nucleocapsid or core, a $T=3$ icosahedral structure composed of 180 subunits of a single polypeptide known as hepatitis B core antigen (HBcAg) (S. Fuller and M. Nassal, personal communication). Within the core is the viral genome, partially duplex 3.2-kb relaxed circular DNA molecule, and a polymerase centrally involved in its replication (ROBINSON and GREENMAN 1974; SUMMERS et al. 1975). Titres of Dane particles in the circulation are extremely variable from individual to individual (and often within a single individual over time), ranging from 10^3 to 10^9 per milliliter of blood.

The 20-nm particles initially visualized in HBV preparations are now recognized to be subviral particles that are composed solely of HBsAg and host-derived lipid (PETERSON 1981; SHIH and GERIN 1977; AGGERBECK and PETERSON 1985). Their lipids consist mainly of phospholipids (notably phosphatidyl choline), free and esterfied cholesterol, and small quantities of triglycerides (GAVILANES et al. 1982). They contain no viral genetic information and are therefore noninfectious; they also lack nucleocapsid components and have no detectable cellular polypeptides. They are enormously abundant in the bloodstream of infected carriers, typically exceeding Dane particles in number by a factor of 10^3–10^5. Paradoxically, they are also highly immunogenic: in most individuals they elicit a strong antibody response that is capable of conferring immunity to reinfection in vivo. For this reason, 20-nm particles purified from infected serum (or generated by recombinant DNA technology, see below) have been exploited as a subunit vaccine for HBV infection (SZMUNESS et al. 1980; VALENZUELA et al.1982; MCALEER et al. 1984; WAMPLER et al. 1985).

The role of these particles in the natural history of infection is unknown, but an attractive speculation is that they serve to adsorb neutralizing antibodies and thus facilitate the progression of infection by virions. That such adsorption occurs in vivo is clearly attested to by the frequent development of arthritis due to the deposition of HBsAg/anti-HBs immune complexes in the joints of acutely

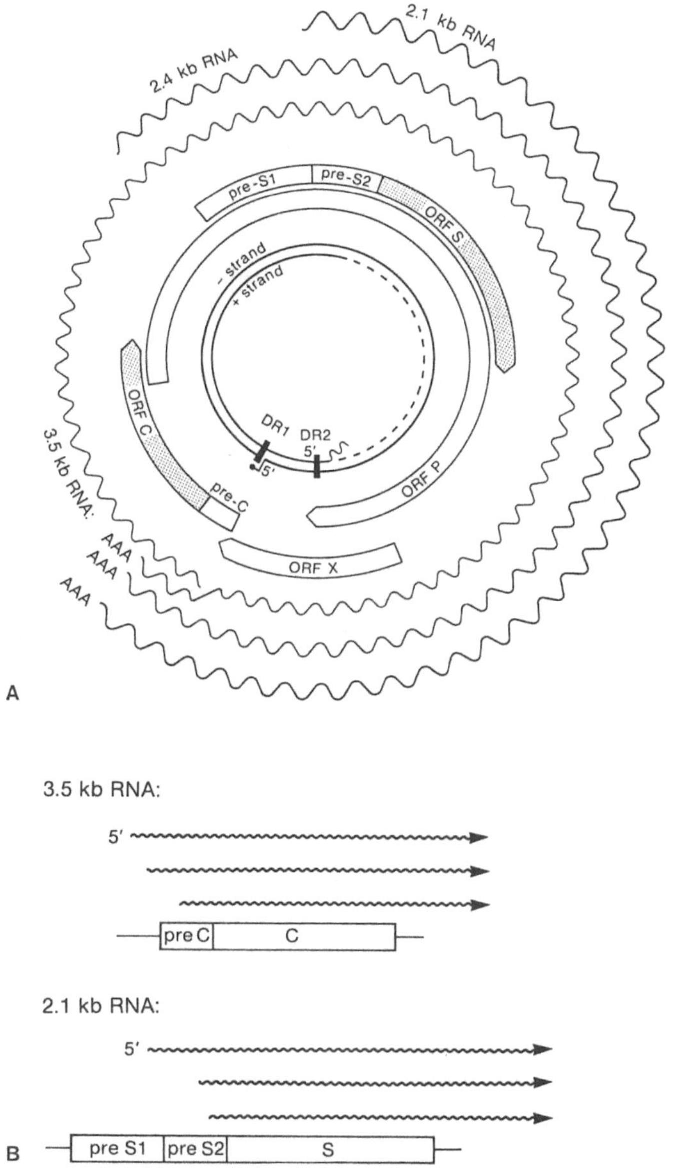

Fig. 1. A The HBV genome. *Wavy lines*, viral transcripts; *boxes*, viral open reading frames (ORFs); *arrows*, direction of transcription and translation; *innermost circles*, structure of virion DNA; *DR1*, *DR2*, positions of the direct repeat sequences involved in the priming steps in viral DNA synthesis. **B** Fine structure of the 5′ ends of the core (*top*) and 2.1-kb surface (*bottom*) mRNAs, relative to the translation starts of the preC and C coding regions (*top*), and preS2 and S coding regions (*bottom*). (From GANEM and VARMUS 1987)

infected patients (GOCKE 1975). What remains at issue is whether these complexes do indeed shield Dane particles from neutralization and accelerate horizontal spread within the liver. One way to approach this problem would be to examine viral mutants that can synthesize virions but not subviral particles. Unfortunately, attempts to generate such mutants have thus far been unsuccessful (V. BRUSS and D. GANEM, unpublished results), a fact that points out both the rudimentary state of our knowledge of the pathways of viral assembly and the potential utility of a deeper understanding of this area.

Since HBV morphogenesis is closely coupled to other aspects of viral replication, such an understanding requires an appreciation of the overall life cycle of hepadnaviruses. Detailed analyses of each of the major steps in viral replication can be found in other chapters in this volume (see also GANEM and VARMUS 1987); here I provide only a brief overview of the events in which viral assembly is embedded. Following the uptake of virions, viral nucleocapsids are transported to the cell nucleus, where their genomes are converted to a covalently closed circular (CCC) form believed to serve as the template for transcription by host RNA polymerase II. Transcription of the genome results in the formation of both subgenomic and genomic mRNAs whose structure is depicted in Fig. 1. The 2.1- and 2.4-kb transcripts are the mRNAs for the viral surface antigens (CATTANEO et al. 1983; STANDRING et al. 1984; WILL et al. 1987). The genomic RNAs are bifunctional, serving not only as the mRNAs for the core proteins (and for the polymerase) but also as the templates for reverse transcription back into the DNA genome (SUMMERS and MASON 1982; ENDERS et al. 1985; BÜSCHER et al. 1985; WILL et al. 1987). Following their transport to and translation in the cytoplasm, the first major event in virion assembly occurs: the selective encapsidation of genomic RNA into the nucleocapsid, along with the viral polymerase. Once encapsidation is underway or completed, reverse transcription of the RNA begins, generating first minus strand, then partially duplex DNA intermediates (SUMMERS and MASON 1982; LIEN et al. 1986; SEEGER et al. 1986). Nucleocapsids containing completed DNA genomes then acquire their outer envelope, presumably by budding into the endoplasmic reticulum (ER) in areas derivatized by the insertion of transmembrane surface antigen polypeptides. The resulting Dane particles are then transported from the cell by the normal pathways of vesicular transport. Viral assembly and release thus occur without cell lysis. In fact, there is no evidence that any of the steps in viral replication is injurious to the host cell, and cell lines that stably produce HBV grow normally in culture (SUREAU et al. 1986; TSURIMOTO et al. 1987; SELLS et al. 1987).

2 Core Particle Formation

Nucleocapsid formation involves the correct assembly of three viral constituents: core antigen, the polymerase and the genomic RNA template. No major cellular proteins have been detected in the core but a role for host factors in nucleocapsid assembly has by no means been excluded. In this section we examine how each of these viral components participates in the assembly process.

2.1 Core Antigens

Core polypeptides are encoded by the C open reading frame (ORF C, Fig. 1; BURRELL et al. 1979; PASEK et al. 1978). Examination of the sequence of this coding region reveals the existence of two in-frame initiation codons (GALIBERT et al. 1979; VALENZUELA et al. 1980). The major nucleocapsid protein originates from the second of these AUGs. To allow ribosomes access to this internal AUG, the viral transcriptional program specifies multiple 5′ ends for the genomic RNAs that serve as the C gene messages (cf. Fig. 1B; ENDERS et al. 1985; WILL et al. 1987; YAGINUMA et al. 1987). The most abundant of these begins between the two initiators, thereby positioning the internal AUG as the first initiator on the transcript, a configuration favorable for efficient initiation of translation (KOZAK 1989). Heterologous viral vectors expressing ORF C from this AUG codon produce a cytoplasmic 21-kDa protein that comigrates with the authentic polypeptide purified from serum-derived virions (WILL et al. 1984; ROOSINCK et al. 1986; MCLACHLAN et al. 1987).

Core antigen has a number of properties in common with the nucleocapsid proteins of other enveloped viruses. First, it is a phosphoprotein: incubation of partially purified cores with gamma-[32P]ATP leads to its phosphorylation on serine residues (ALBIN and ROBINSON 1980; GERLICH et al. 1982; FEITELSON et al. 1982). Similar labeling occurs when transfected cells expressing HBcAg are metabolically labeled with [^{32}P]-orthophosphate (ROOSINCK and SIDDIQUI 1987). In duck HBV (DHBV), and most likely in HBV as well, most of the phosphorylation is at the C-terminus of the protein (SCHLICHT et al. 1989). The predicted amino acid sequence of core antigen does not contain sequences homologous to the kinase domains of known serine/threonine kinases, and HBcAg produced in *Escherichia coli* is unphosphorylated, arguing that the activity is not intrinsic to the C gene product itself (SCHLICHT et al. 1989). Most evidence suggests that the responsible kinase is of host origin, although its precise identity is uncertain. The function of core phosphorylation, if any, is uncertain, but it is noteworthy that this feature has been conserved in all of the animal hepadnaviruses (FEITELSON et al. 1982; SCHLICHT et al. 1989; PUGH et al. 1989). A recent examination of DHBV infection noted that cores from serum-derived virions were underphosphorylated relative to cytoplasmic cores purified from infected cells (PUGH et al. 1989); this raises the interesting possibility that core dephosphorylation might be involved in signaling mature cores for envelopment or export.

Another striking feature of HBcAg is that the C-terminal portion of the molecule is extremely basic, with arginines heavily overrepresented. This suggested early on that the protein would display nonspecific nucleic acid binding activity, an inference that has been sustained by a variety of assays using natural or recombinant HBcAg and labeled DNA or RNA (PETIT and PILLOT 1985; GALLINA et al. 1989). Removal of the C-terminal region abolishes this binding (GALLINA et al. 1989), although a report to the contrary (MATSUDA et al. 1988) has appeared. Heterologous nucleic acids are bound as well as homologous ones, and there is no evidence as yet for sequence specificity in the protein: nucleic acid interactions being studied in these assays. In recent experiments with DHBV, deletions in the C-terminus of the core protein affected neither particle formation nor genomic RNA packaging, but the resulting particles were noninfectious (SCHLICHT et al. 1989).

This is not to say that core polypeptides might not play a role in the selective encapsidation of genomic RNA during nucleocapsid assembly. Such is the case for retroviral *gag* proteins (MERIC and SPAHR 1986), though there is no conservation in hepadnaviruses of the reiterated *cys-his* sequence motifs implicated in this function in retroviruses. Support for such a role, however, must await the development of binding assays in which appropriate target sequences of HBV RNA (see below) are selectively bound in vitro, or the identification of mutants in ORF C that are assembly competent but encapsidation defective. No such mutants have yet been generated.

Core polypeptides are able to assemble into nucleocapsids in the absence of all other viral proteins. This is most clearly demonstrated by the observation that recombinant vectors expressing ORF C in hosts as divergent as COS cells, yeast, and *Escherichia coli* all direct the formation of morphologically correct nucleocapsids (COHEN and RICHMOND 1982; UY et al. 1986; MIYANOHARA et al. 1986; OU et al. 1986). Such assembly is unimpeded by the addition of heterologous sequences to the N-terminus of the chain (CLARKE et al. 1987; STAHL and MURRAY 1989). From the results in *E. coli* it is sometimes inferred that no specialized host proteins are also required, but a role for evolutionarily conserved functions like heat shock proteins or chaperonins—known to be important in other protein oligomerization processes (HEMMINGSEN et al. 1988)—is not excluded.

Another important implication of these results is that pregenomic RNA is not required to nucleate proper core assembly, since this transcript is not present in these systems. Interestingly, nuclease protection experiments indicate that the *E. coli*-produced cores are not entirely devoid of nucleic acid. Rather, they appear to contain moderate amounts of RNA that may be enriched for sequences of viral origin (BIRNBAUM and NASSAL 1990; V. Bruss, personal communication). The significance of this observation is uncertain; the suggested enrichment could simply be due to the fact that HBcAg mRNA may be the most abundant transcript in the cells at the time of assembly. Certainly other viral proteins known to play a key role in encapsidation (see below) are absent in these systems, and conclusions about the role of core polypeptides in genomic RNA selection should be drawn from these experiments with caution.

One last feature of ORF C deserves mention. Recall that this coding region contains two in-frame AUG codons, with the second one used for HBcAg expression. It turns out that the first initiator is also used for translational initiation in vivo, with the longer genomic transcripts that retain this AUG serving as message (Fig. 1B). As OU et al. (1986) were the first to point out, expression from this initiator confers new biologic properties on the polypeptide. The translation of the so-called preC region between the AUGs generates a signal sequence for protein translocation into the ER. From there, the chains travel through the secretory pathway, undergoing cleavage of their C-terminal basic domain in the process. This generates a series of secreted 16–18kDa proteins with altered antigenicity, now termed hepatitis B e antigen (HBeAg; OU et al. 1986; MCLACHLAN et al. 1987; STANDRING et al. 1988). In infected patients these proteins can be detected in the bloodstream; where they are correlated with elevated titres of circulating Dane particles and correspondingly high infectivity (OKADA et al. 1976; GANEM 1982).

The processing protease is likely a cellular enzyme that is (partially) inhibited by high concentrations of pepstatin, suggesting it may be an aspartyl protease (JEAN-JEAN et al. 1989a). Suggestions that the protease is of viral origin, stemming from the recognition of regions of sequence homology between the N-terminal region of ORF C and known eukaryotic aspartyl proteases (MILLER 1987), have not been supported by genetic studies: mutational ablation of the putative active sites of this protease does not impair the processing (NASSAL et al. 1989; JEAN-JEAN et al. 1989b).

The ER translocation reaction displays several unusual properties, one of which is its inefficiency: many pre-C chains of woodchuck hepatitis virus (WHV) are apparently not translocated at all, judging from the failure of their signal sequence to be removed (WEIMER et al. 1987). This is highly unusual, since for most secreted proteins nascent chains are cotranslationally translocated across the ER bilayer and their signal sequences removed even before synthesis of the polypeptide is completed. Even more remarkable is the observation that, following partial translocation into the ER lumen and cleavage of the signal sequence by signal peptidase, chain translocation is frequently aborted and the N-terminally processed chain "regurgitated" back into the cytoplasm (GARCIA et al. 1988). Strikingly, recent experiments of OU et al. (1989) indicate that many of the processed chains can also be transported to the nucleus! It is unclear whether the nuclear material derives from the translocated or the regurgitated population of processed chains, though the latter seems more likely. Studies of deletion mutants in the preC region implicate the ten preC amino acid residues between the signal peptide cleavage site and the C AUG as essential to nuclear transport. The important nuclear translocation signals reside in the basic C-terminus of the core proteins; perhaps these preC sequences affect the conformation or accessibility of these signals (YEH et al. 1990). These experiments help to explain the long-standing observation of the nuclear localization of HBcAg reactivity in immunofluorescence studies (HOOFNAGLE et al. 1975, 1978). The function of this nuclear transport reaction in the life cycle is unproven, though it is tempting to believe that it has something to do with the nuclear delivery of encapsidated DNA genomes for conversion to CCC

DNA (TUTTLEMAN et al. 1986). However, this will require proof that the regurgitated form of core antigen (a) is assembled into nucleocapsids containing viral DNA; and (b) can serve as the precursor of the nuclear HBcAg. In addition, the fact that mutants deficient in this protein are viable (see below) means that this cannot be the only nuclear delivery pathway for cores.

The observation that preC translation targeted C-related proteins to the ER, together with the known clinical correlation of serum HBeAg with infectivity, initially suggested that precore proteins might play an essential role in virion morphogenesis (OU et al. 1986; BRUSS and GERLICH 1988). The fact that the preC region is conserved in all animal hepadnaviruses (KODAMA et al. 1985) made this speculation still more attractive. However, mutational ablation of preC translation in DHBV was soon shown to have no effect on viral replication, assembly, and infectivity in both cell culture and the intact animal host (CHANG et al. 1987; SCHLICHT et al. 1987b). Subsequent studies have detected naturally occurring HBV preC-deficient mutants in the circulation of some HBV carriers as well (BRUNETTO et al. 1989; OKAMOTO et al. 1990). To this day, the function of e antigen remains in enigma. SCHLICHT and SCHALLER (1989) have recently shown that HBeAg determinants can be expressed on the surface of cells infected with recombinant vaccinia vectors expressing this protein, and that this expression does not result from the nonspecific adsorption of secreted HBeAg to the cell membrane. Perhaps it is delivered to the surface by pathways akin to those involved in the presentation of endogenous antigens to cytotoxic T cells, via an association of peptide fragments with class I major histocompatibility (MHC) molecules (TOWNSEND and BODMER 1989). If so, then e antigen may be important in triggering cytotoxic host immune responses to HBV that may play a role in disease pathogenesis or viral clearance. But these activities would be expected to restrict or curtail virus spread. So the paradox remains: Why has expression of pre-C been retained in viral evolution?

2.2 Encapsidation of Genomic RNA

The assembly of a complete nucleocapsid requires that the pregenomic RNA that is the template for reverse transcription be packaged, along with the viral polymerase. RNA packaging is selective: detailed studies of the viral RNA within cores of ground squirrel hepatitis virus (GSHV) and DHBV indicates that genomic but not subgenomic RNAs are encapsidated (ENDERS et al. 1987; HIRSCH et al. 1990). The mechanism of selection appears to be stringent. Even among the family of genomic RNAs, whose members differ by only 31 nt owing to the heterogeneous placement of the 5′ end (cf. Fig. 1B), only a single RNA—the shortest—is packaged. Since the longer genomic RNAs contain all of the sequences present in the shortest, it is clear that sequence alone cannot entirely account for this discrimination. Recent studies from two groups have shed light on this interesting step in assembly, with regard to both the proteins involved in RNA selection and the nature of the *cis*-acting recognition elements.

In both DHBV (HIRSCH et al. 1990) and HBV (BARTENSCHLAGER et al. 1990; R. HIRSCH, unpublished data), viruses with null mutations in the polymerase gene (ORFP) are unable to encapsidate genomic RNA, producing viral cores with no viral RNA. In the DHBV case, the inability to package genomic transcripts did not lead to enhanced encapsidation of other viral RNAs, as might have been expected if core antigen had a limited ability to carry out this step. The results indicate that the P protein has an essential role in the encapsidation reaction, though the mechanistic details of this role are unknown. Among the possibilities are that the protein interacts directly with specific sequences on the RNA, or that it forms a noncovalent complex with other proteins (e.g., core antigen) which then allows binding of the complex to the transcript; alternatively, the polymerase might initiate DNA synthesis on free pregenomic RNA, with the nascent RNA:DNA hybrid forming the packaging signal that is recognized by core proteins. Of these models, only the last can be excluded at present, since missense mutations in the P gene that destroy the ability of the resulting protein to carry out DNA synthesis nonetheless package RNA at normal efficiency (HIRSCH et al. 1990; BARTENSCHLAGER et al. 1990).

A curious feature of the involvement of P gene products in packaging is that it is *cis*-preferential. That is, in most cases in which wild-type genomes are used to complement P^- mutants, the encapsidated progeny genomes are predominantly of wild-type origin (LAVINE et al. 1989; HIRSCH et al. 1990; BARTENSCHLAGER et al. 1990). The mechanism underlying this is not understood. It could simply reflect the limiting quantities of P proteins in the cell, leading to a preference for action on RNAs nearest to their site of synthesis; alternatively, there could be a preference for nascent P chains to interact with their own genomic mRNA while still on polysomes. However, *cis*-packaging is clearly not obligatory. A low level of *trans*-complementation does occurs, and recombinant vectors designed to overexpress P proteins complement P^- mutants substantially more efficiently (T. WU, L. CONDREAY, and W. MASON, unpublished results; BARTENSCHLAGER et al. 1990). This indicate that there is no mechanistic barrier to RNA recognition and packaging by mature released chains.

Similarly rapid progress has been made in defining the *cis*-acting elements that earmark the genome for encapsidation. JUNKER-NIEPMANN et al. (1990) have shown that deletions originating within the preC region of HBV are packaging defective, and that transfer of sequences spanning this region to the 5′ end of a heterologous (lacZ) mRNA allowed encapsidation of the latter in HBV cores. Likewise, in the DHBV system, preC linker insertions and deletions abolish encapsidation (R. HIRSCH and D. GANEM, submitted). The involved regions of both viruses are present in the terminal redundancy of the genomic RNA, raising the question of which of the repeated copies is recognized by the packaging machinery. The study of DHBV mutants in which the 5′ or 3′ copy of this sequence was selectively inactivated has recently shown that only the 5′ copy is active in packaging: deletion of the 3′ copy allows packaging, while deletion of the 5′ copy inactivates pregenome encapsidation (R. HIRSCH and D. GANEM, submitted). This is in contrast to the nearby sequences that specify the initiation of minus strand DNA synthesis, which (in WHV, at least) are only active in the 3′ position (SEEGER and MARAGOS 1990).

The location of the sequences that mediate RNA packaging does not explain the discrimination of the packaging machinary against the longer species of genomic RNAs, since the packaging signal is present in all three species. As noted previously, some other feature is required to explain this discrimination. Early studies (ENDERS et al. 1987) raised several possible models: e.g., that a secondary structure unique to pregenomic RNA allowed access to the packaging signal; that a specific relationship to the 5′ cap was necessary; or that translation of the preC region, which is only allowed by the longer transcripts containing the preC AUG, might sequester them from the packaging apparatus. Recently, NASSAL et al. (1990) have presented strong evidence in favor of the translation model, by showing that encapsidation of longer genomic RNAs occurs when the potential for preC translation is removed by mutation.

3 Subviral HBsAg Particle Formation

As noted previously, a hallmark of HBV infection is the overproduction of subviral particles composed only of the viral surface proteins and lipid. Although early immunologic studies suggested that only one viral antigen was present on these particles (and on virions), we now know that the repertoire of envelope proteins produced by the virus is more complex. This complexity, in turn, provides the diversity of structural elements that is required to build different structures out of shared components.

3.1 The Surface Antigen Family

The predominant viral surface protein in the circulation of infected individuals is the 24-kDa S protein, which (together with its glycosylated form) is present in concentrations of up to 50–300 μg/ml. The coding region for this protein was identified by aligning the translated HBV DNA sequence with the known N-terminal amino acid sequence of the S chain (VALENZUELA et al. 1979). This immediately revealed that the open reading frame for the S antigen was open for some 400 nt 5′ to the S AUG, and that this upstream (or preS) ORF could be divided into two subregions (preS1 and preS2) by the presence of two in-frame initiation codons (GALIBERT et al. 1979; VALENZUELA et al. 1980; see Fig. 1). Thus, the sequence raised the possibility that two larger, S-related proteins could be generated during infection by translation of this preS region. The preservation of this coding scheme in all the mammalian hepadnaviruses further supported the potential functional importance of this arrangement (GALIBERT et al. 1982; SEEGER et al. 1984). It was not long before both polypeptides were detected in all three mammalian hepadnaviruses (HEERMAN et al. 1984; WONG et al. 1985;

PERSING et al. 1986). The avian hepadnaviruses also contain a preS region with multiple in-frame AUGs, but only one preS protein has thus far been reproducibly detected (PUGH et al. 1987; SCHLICHT et al. 1987a).

In HBV, the largest of these so-called preS proteins is the 39-kDa L protein, which is the product of initiation at the first AUG of the ORF (HEERMAN et al. 1984). Initiation at the second AUG generates the 31-kDa M protein (MACHIDA et al. 1983, 1984; STIBBE and GERLICH 1983; PERSING et al. 1985). Both proteins share the common C-terminal S domain and differ principally by the length and structure of their N-terminal extensions. Both are quantitatively minor components of the circulating pool of S-related antigens, with the M protein accounting for approximately 5%–15% of the total, and the L protein typically representing only 1%–2% (or less) of the total. Each protein exists in two isomeric forms differing only by the presence or absence of S domain glycosylation; in addition, the M protein contains another N-linked oligosaccharide on its preS2-specific domain. The L protein contains an additional post-translational modification, a myristic acid group in amide linkage to its amino terminal glycine residue (PERSING et al. 1987).

As in the C coding region, the problem of ribosomal access to the internal AUGs is solved by the deployment of separate transcripts immediately 5′ to each AUG. In this case two promoters are required: the upstream promoter (P1) specifying the L mRNA, and the downstream (P2) specifying the M and S mRNAs (WILL et al. 1987; YAGINUMA et al. 1987). As might be expected from this arrangement, no precursor-product relationship exists between the preS and S proteins: frameshift or deletion mutations within preS do not affect S biosynthesis (PERSING et al. 1985). The relative levels of these transcripts in infected cells probably accounts for most of the differences in the stoichiometry of the circulating proteins. The L mRNA is present in very low quantities in infected liver (WILL et al. 1987), corresponding to its low concentration in serum, while S message is highly abundant. The case of the M protein requires further examination. Nuclease mapping studies show that M message can be present at levels of up to 30%–50% that of S mRNA (CATTANEO et al. 1983; STANDRING et al. 1984), yet the relative levels of M to S protein are usually measurably less than this. Since the stabilities of these two proteins are not markedly different, this may indicate a modest role for translational control in M protein expression.

3.2 The S Protein and the Assembly of 20-nm Particles

The 20 nm particles purified from carrier serum typically contain approximately 100 polypeptide subunits per particle, predominantly the S protein with variable amounts of M and only trace quantities of L subunits (HEERMAN et al. 1984). With the advent of cloned HBV DNA, it became possible to express the surface protein coding regions in heterologous cells, leading to the discovery that cells expressing M and S subunits assemble and secrete 20-nm particles that are indistinguishable from those found in serum (DUBOIS et al. 1980). This observation forms the basis of all current strategies for the production of recombinant BHV vaccines (see below). Cells expressing only the S subunit are also capable of producing morphological normal 20-nm particles, indicating that all of the information necessary for this assembly process resides in the S domain (LIU et al. 1982; LAUB et al. 1983). The cellular machinery involved in particle assembly is not limited to differentiated hepatocytes: virtually all higher eukaryotic cells that express the protein will assemble and secrete 20-nm particles (MORIARTY et al. 1981; STENLUND et al. 1983; STRATOWA et al. 1982). Insect cells (LANFORD et al. 1989) and *Xenopus* oocytes (SIMON et al. 1988b), however, carry out these processes very inefficiently (see below). Yeast cells do not secrete HBsAg at all, but lipoprotein arrays of S subunits do form intracellularly, and such complexes have been successfully purified and used as a vaccine (VALENZUELA et al. 1982; WAMPLER et al. 1985).

The pathway of 20-nm particle assembly is now understood in some detail. Early morphologic studies of infected liver revealed that particles accumulate within cisternae of endoplasmic reticulum, indicating that particle formation occurs intracellularly in membrane-limited vesicles of the secretory pathway (GERBER et al. 1974). These observations have been sustained by more recent electron-microscopic studies of transfected, HBsAg-positive cells in culture (PATZER et al. 1986) and are consistent with the fact that the chains acquire asparagine-linked high-mannose oligosaccharide chains, a modification that occurs only in the ER. Assembled complexes are transported through the constitutive secretory pathway (KELLY 1985). En route to the exterior they traverse the Golgi complex, where their high mannose carbohydrates are processed to the complex, endoglycosidase H-resistant form (PATZER et al. 1984). Transfer of the complexes from the ER to the Golgi appears to be the rate-limiting step in export, since all intracellular S antigen is found in the endoH-sensitive form, whereas all extracellular antigen is endoH-resistant.

How are particles formed in the ER? The first clues came from studies of S protein processing in cell-free translation systems supplemented with ER-derived microsomal vesicles (EBLE et al. 1986, 1987). In these experiments, nascent S protein chains were found to be inserted into the ER bilayer as transmembrane polypeptides, as judged by their partial resistance to exogenous proteases and their ability to remain associated with the bilayer at high pH. The protein is oriented in the bilayer such that both N- and C-termini project into the vesicle lumen; to do so, the polypeptide must span the bilayer at least twice. This topology is achieved through the conjoint action of two complex signal sequences (EBLE et al. 1987). Signal I is an N-terminal signal sequence (residues 4–28) that initiates chain translocation across the bilayer; it resembles conventional signal sequences save for the fact that it remains uncleaved during translocation. Signal II (residues 80–100) is a complex topogenic element that serves both as a stop-transfer (anchoring the protein in the bilayer with residue 80 facing the cytoplasm and residue 100 facing the ER lumen) and a signal sequence directing the lumenal translocation of the C-terminal region of the molecule. In these regards, signal II resembles the internal signal/stop transfer elements of the asialoglycoprotein and transferrin receptors (SPEISS and LODISH 1986).

In the in vitro translocation system, the transmembrane form of the S protein is the final product (EBLE et al. 1986). However, in vivo this form is only a transient intermediate in particle formation. Pulse-chase experiments in mouse L cells expressing the S protein reveal that transmembrane forms identical to those observed in vitro are the initial product of synthesis but are rapidly converted to a series of defined assembly intermediates (SIMON et al. 1988a). Following their insertion into the bilayer, the transmembrane monomers aggregate and host proteins are excluded from the complex. The aggregate is then transposed into the lumen by a process of extrusion or budding. It is likely that this extrusion event differs from other budding reactions in that the "budded" product lacks a typical unit membrane structure and has a lipid content much lower than that of most cellular membranes; this suggests that the host lipids may be substantially reorganized in the process. The details of this interesting reaction, however, are unknown.

The pathway outlined above bears many features in common with morphogenetic events postulated for the envelope proteins of other enveloped viruses. Chief among these are insertion as a transmembrane polypeptide, the involvement in lateral interactions with other viral glycoproteins in the plane of the bilayers, and the exclusion of host proteins. However, in other systems, envelope glycoproteins are not released as such from infected cells. Instead, they remain moored in the bilayer until interactions with other viral components (nucleocapsid and/or matrix proteins) trigger the budding step. This ensures that envelope glycoproteins never exist the cell without enveloping the nucleocapsid. The distinctive feature of the S protein is that it can carry out the entire assembly sequence without the participation of other viral proteins. This results in the release of the subviral particle containing only surface proteins that is the hallmark of hepadnaviral infection.

Coexpression of M and S polypeptides in the same cell results in the formation of mixed particles containing both species. The mixed particles are morphologically identical to pure S particles and have the same buoyant density. They can be distinguished by their reactivity with antisera specific for the preS2 domain of M and by their ability to bind polymers of human albumin, another activity mediated by this domain (MICHEL et al. 1984; MACHIDA et al. 1984; PERSING et al. 1985). (The biologic function of polyalbumin binding, if any, is unknown). These data indicate that the preS2 domain is exposed on the surface of the particle. Given current views of particle formation, this would suggest that in the transmembrane form of the M protein, this domain must be translocated into the ER lumen. Recent experiments in the cell-free translocation system have confirmed this, and demonstrated in addition that the preS2 region does not contain its own signal sequence activity (EBLE et al. 1990). This means that its transmembrane transport must be mediated by the *downstream* signals in the S domain (e.g., signal I). This unusual behavior had previously only been identified in experimental fusion proteins constructed in vitro (PERARA et al. 1986).

Mixed M/S particles are efficiently secreted by most cells (MICHEL et al. 1984; PERSING et al. 1985; CHENG and MOSS 1987). Studies reporting that M protein overexpression is inhibitory to S export (STANDRING et al. 1986) were conducted in *Xenopus* oocytes, where the efficiency of S export is already very low; inhibitory effects do not occur in mammalian cells. Particles containing both M and S subunits are of great interest as potential vaccines because of their enhanced immunogenicity, as documented in immunization studies of inbred mice. Even mice genetically incapable of responding to S determinants will make an anti-S response following immunization with such particles (MILICH et al. 1985). It is hoped that similar results can be achieved in the appreciable number of humans who respond poorly or not at all to S protein-based vaccines (ALPER et al. 1989).

Recent studies have sought to define subdomains of the S protein that might mediate one or more of the steps in S particle assembly (V. BRUSS and D. GANEM, in preparation). While simple in principle, such experiments are bedeviled in practice by the unpredictable effects of mutations (especially deletions) on protein folding and stability. Some conclusions can nonetheless be made. First, insertions into one region of the cytoplasmically disposed loop region between signals I and II are generally well tolerated, with minimal effects on particle assembly and secretion (DELPEYROUX et al. 1986, 1987; V. BRUSS and D. GANEM, unpublished results). Lesions affecting the proper disposition of the N-terminus (e.g., deletions of signal I, substitution of cleaved signals for signal I, or addition of bulky extensions to the N-terminus) are poorly tolerated; although they enter the secretory pathway (as judged by glycosylation), they fail to assemble either with each other, or with wild-type S chains when these are coexpressed with them. Interestingly, the ability of one such molecule (a globin–S fusion protein in which approximately 100 aa of alpha globin were added to the N-terminus of S) to be efficiently assembled and secreted was rescued by restoring its correct transmembrane topology through provision of an N-terminal signal sequence (V. BRUSS and D. GANEM, in preparation). These data attest to the importance of the proper orientation of the first portions of the S chain in allowing lateral interactions with other monomers. Consistent with this, an amber mutant missing the C-terminal hydrophobic domain of S remains competent to aggregate with other S chains, since it is efficiently cosecreted with wild-type S chain when the two are coexpressed. Clearly, further understanding of the details of S particle assembly will require both more refined genetic studies and the development of an in vitro system in which correct particle formation can be observed. One useful byproduct of such an understanding could be the ability to rationally design chimeric particles containing foreign epitopes for possible use as vaccines. Several successful chimeras have already been generated empirically (DELPEYROUX et al. 1986; 1987; VALENZUELA et al. 1985; BRUSS and GANEM 1991), but progress has been slow due to both the lack of precise information about S assembly domains and to the twin problems of misfolding and instability mentioned above.

3.3 The L Protein and 20-nm Particle Formation

We have previously noted that natural 20-nm spherical particles often lack detectable L protein, and that the L protein is overall a minor component of the circulating pool of surface antigens (HEERMAN et al. 1984). The reason for the low abundance of L chains on 20-nm particles became clearer in experiments aimed at overexpression of ths subunit in cultured cells. Vaccinia vectors that express only the L protein revealed that the L chains themselves are not secreted, despite the presence of an intact S domain that contains all the information required for this process (CHENG et al. 1986). This suggests that the preS1-specific domain of L contains signals that promote intracellular retention of the chain and that can override the secretory elements in the S domain. While the topology of this protein was not examined in these experiments, in other studies in *Xenopus* oocytes we have observed that the protein remains in a transmembrane configuration and does not progress to the later, protease-resistant forms that characterize normal S morphogenesis (K. SIMON and D. GANEM, unpublished results). Epitope mapping studies with preS1-specific monoclonal antibodies suggest that most of the preS1 domain is disposed on the lumenal side of the ER membrane (KUROKI et al. 1990).

When L chains are overexpressed in the presence of M and S chains, an additional finding emerges: that the ability of the S and M chains to be secreted is specifically inhibited in a dose-dependent fashion (CHENG et al. 1986; PERSING et al. 1986; STANDRING et al. 1986; CHISARI et al. 1986). All three chains accumulate intracellularly, and examination of their carbohydrate side chains indicates that they are retained in a pre-Golgi compartment; electron-microscopic studies confirm this to be the endoplasmic reticulum (CHISARI et al. 1987). Presumably this retention reflects the formation of mixed arrays of L-M-S aggregates, with the inhibitory influence of the preS1 domain now being conferred upon the entire aggregate. The inhibitory effect of L overexpression on S particle secretion explains the underrepresentation of L chains on 20-nm spherical particles. Interestingly, in electron-microscopic studies of the hepatocytes of transgenic mice overexpressing L chains (relative to S), 20-nm filamentous forms were frequently observed in the ER (CHISARI et al. 1987). This accords well with earlier observations that filaments are relatively enriched in L subunits (HEERMAN et al. 1984) and suggests that filaments arise in regions of the ER where local concentrations of L protein are high.

The overaccumulation of retained L-M-S aggregates can be injurious to the host cell. Hepatocytes of transgenic mice bearing these aggregates develop characteristic cytopathic effects that over long periods of time provoke hepatocellular regeneration and ultimately hepatocellular carcinoma (CHISARI et al. 1987, 1989). The implications of these findings for viral pathogenesis will be considered elsewhere in this volume. Suffice it to say here that in liver biopsy specimens from cases of acute hepatitis B cells with similar cytopathic effects (so-called ground glass hepatocytes) are not infrequently seen, and it seems likely that these reflect the consequences of L protein overproduction in these cells.

What accounts for the intracellular retention of surface antigens by L chains? Clearly this property is specified, directly or indirectly, by the preS1 domain. Deletion mutagenesis of this region indicates that the inhibitory sequences map to the extreme N-terminus of the molecule (KUROKI et al. 1989). The observation that this region of the L protein is myristylated raised the possibility that ER retention might be mediated by the affinity of this hydrophobic fatty acid for the ER membrane. In other cellular proteins known to harbor this modification it is frequently (but not invariably) involved in promoting associations with cellular membranes (TOWLER et al. 1988). Examinations of the role L protein myristylation in secretion have led to divergent results. Missense mutations that ablate the ability of the L protein of the *adw* serotype of HBV to be myristylated clearly have no effect on secretion-inhibition, indicating that in this serotype primary sequence features play the dominant role (KUROKI et al. 1989). However, similar studies employing the *ayw* serotype suggest a prominent role for myristylation in this activity (R. PRANGE, A. CLEMEN, and R. STREECK, personal communication). These differences are most likely attributable to the fact that the sequences of these two L proteins differ significantly at their N termini, with the *adw* serotype containing an insertion of an added 11 amino acids in this position. This molecule is more potent as an inhibitor of secretion than is its *ayw* counterpart (R. RUSSNAK and D. GANEM, unpublished results), though both are functional in vivo, as judged by the infectivity of their parent clones.

4 Dane Particle Assembly

The extensive progress of the past few years in understanding the biogenesis of subviral HBsAg particles has provided a firmer basis for thinking about the mechanism of Dane particle assembly. Since all three envelope proteins are expressed primarily on intracellular membranes rather than on the cell surface (OU and RUTTER 1987), virion formation almost certainly proceeds on intracellular membranes; electron-microscopic studies consistent with this have been reported (KAMIMURA et al. 1981). Perhaps the most important additional advance in this area was the recognition by Gerlich and colleagues that the envelope polypeptide composition of virions differs significantly from that of 20-nm spheres (HEERMAN et al. 1984). They showed, that unlike the latter, virions contained relatively large quantities of L proteins; in their preparations the ratio of L:M:S subunits was approximately 1:1:4. Since these preparations are unavoidably contaminated to some extent with 20-nm particles, it is likely that the ratio of L:S chains in authentic virions is still higher.

More recent studies have used genetic approaches to define the components required for Dane particle production. Cultured hepatoma cells (HepG2, Huh6, and Huh7) transfected with HBV DNA secrete infectious virions (SUREAU et al. 1986; TSURIMOTO et al. 1987; YAGINUMA et al. 1987; SELLS et al. 1987). By transfecting these cells with missense or amber mutants selectively affecting the L, M, or S proteins, two groups have shown that the L and S proteins are required for correct virion morphogenesis (K. UEDA, T. TSURIMOTO, and K. MATSUBARA, unpublished results; BRUSS and GANEM 1991). (The role of the M protein in assembly is still disputed.) Unlike retroviruses, whose nucleocapsids can still bud out of cells in the absence of envelope proteins, no HBV cores are exported if either of these proteins is absent. HBV missense mutants affecting L myristylation remain competent for virion assembly and secretion (BRUSS and GANEM 1991); although the infectivity of these mutants is difficult to test, analogous mutants of DHBV are known to be noninfectious in the animal host (MACRAE, BRUSS and GANEM 1991). If, as seems likely, the same is true for HBV, then myristylation may play its most important role in the adsorption, entry, or uncoating of virions on the next cycle of infection.

The secretary behavior of the L protein and its enrichment on virions have led to a model for virion assembly (BRUSS and GANEM 1991) that attributes a central role in this process to the L chain. In this view, L is the analog of the conventional envelope glycoproteins of other viruses (e.g., VSV G or influenza HA), moored in the membrane awaiting interactions with nucleocapsids. In this view, the interaction of such chains with cores would precipitate the same series of morphogenetic events that for the S proteins occurs constitutively, perhaps by triggering a conformational change in the L polypeptide. The genetic data indicate that the S proteins must also play an active role in the budding process, since L proteins cannot perform this act alone. Perhaps the precise stoichiometry of the L and S chains is important; suboptimal ratios may lead instead to filament formation.

Clearly, much remains to be explained. How do cores trigger budding? Which regions of core antigen interact with which regions of the surface protein(s)? Why do only cores containing mature (replicated) genomes appear extracellularly? Do additional morphologic changes occur distal to the budding step, as virions are transported through the secretory pathway? These and other lingering questions will surely preoccupy students of hepadnaviral morphogenesis for many years to come.

References

Aggerbeck LP, Peterson DL (1985) Electron microscopic and solution X-ray scattering observations on the structure of hepatitis B surface antigen. Virology 141: 155–161

Albin C, Robinson WS (1980) Protein kinase activity in hepatitis B virions. J Virol 34: 297–302

Alper CA, Kruskall MS, Marcus-Bagley D, Craven DE, Katz AJ, Brink SJ, Dienstag JL, Swdeh Z, Yunis EJ (1989) Genetic prediction of nonresponse to hepatitis B vaccine, N Engl J Med 321: 708–712

Bartenschlager R, Schaller H (1988) The amino terminal domain of the hepadnaviral P gene encodes the terminal protein (genome-linked protein) believed to prime reverse transcription. EMBO J: 4185–4192
Bayer M, Blumberg B. Werner B (1968) Particles associated with Australia antigen in the sera of patients with leukemia, Down's Syndrome and hepatitis. Nature 218: 1057–1059
Birnbaum F and Nassal M (1990) Hepatitis B virus nucleocapsid assembly: Primary structure requirements in the core protein J Virol 64: 3319–3330
Bartenschlager R, Junker-Niepmann M and Schauer H (1990) The P protein of HBV is required as a structural component for genomic RNA encapsidation J Virol 64: 5324–5332
Blumberg BS, Alter HJ, Visnich (1965) A "new" antigen in leukemia sera. JAMA 191: 541–546
Brunetto M, Stemmler M, Schodel F, Will H, Ottobrelli A, Rizetto M, Verme, Bonino F (1989) Identification of HBV variants which cannot produce precore-derived HBeAg and may be responsible for severe hepatitis. Ital J Gastroenterol 21: 151–154
Bruss V and Ganem D (1991) The role of envelope proteins in hepatitis B virus assembly. Proc Nat Acd Sci USA 88: 1059–1063
Bruss V, Gerlich WH (1988) Formation of transmembranes hepatitis B e-antigen by cotranslational *in vitro* processing of the viral precore protein. Virology 163: 268–275
Burrell CJ, Mackay P, Greenway PJ, Hofschneider PH, Murray K (1979) Expression in *Escherichia coli* of hepatitis B virus DNA sequences cloned in plasmid pBR322. Nature 279: 43–47
Büscher M, Reiser W, Will H, Schaller H (1985) Transcripts and the putative RNA pregenome of duck hepatitis B virus: implications for reverse transcription. Cell 40: 717–724
Cattaneo R, Will H, Hernandez N, Schaller H (1983) Signals regulating hepatitis B surface antigen transcription. Nature 305: 336–338
Chang C, Enders G, Sprengel R, Peters N, Varmus HE, Ganem D (1987) Expression of the precore region of an avian hepatitis B virus is not required for viral replication. J Virol 61: 3322–3325
Cheng C, Smith K, Moss B (1986) Hepatitis large surface protein is not secreted but is immunogenic when selectively expressed by recombinant vaccinia virus. J Virol 60: 337–344
Cheng KC, Moss B (1987) Selective synthesis and secretion of particles composed of the hepatitis B virus middle surface protein directed by a recombinant vaccinia virus: induction of antibodies to preS and S epitopes. J Virol 61: 1286–1290
Chisari FV, Fillipi P, MacLachlan A, Milich D, Riggs M, Lee S, Palmiter R, Pinkert C, Brinster R (1986) Expression of hepatitis B large envelope polypeptide inhibits hepatitis B surface antigen secretion in transgenic mice. J Virol 60: 880–887
Chisari FV, Filippi P, Buras J, MacLachlan A, Popper H, Pinkert C, Palmiter R, Brinster R (1987) Structural and pathological effects of synthesis of hepatitis B virus large envelope polypeptide in transgenic mice. Proc Natl Acad Sci USA 84: 6909–6913
Chrisari FV, Klopchin K, Moriyama T, Pasquinelli C, Dunsford HA, Brinster RL, Palmiter RD (1989) Molecular pathogenesis of hepatocellular carcinoma in hepatitis B virus transgenic mice. Cell 59: 1145–1156
Clarke BE, Newton SE, Carroll AR, Francis MJ, Appleyard G, Syred AD, Highfield PE, Rowlands DJ, Brown F (1987) Improved immunogenicity of a peptide epitope after fusion to hepatitis B core protein. Nature 330: 381–384
Cohen B, Richmond J (1982) Electron microscopy of hepatitis B core antigen synthesized in *C. coli*. Nature 296: 677–678
Dane DS, Cameron CH, Briggs M (1970) Virus-like particles in serum patients with Australia antigen-associated hepatitis. Lancet 1: 695–698
Delpeyroux F, Chenciner N, Lim A, Malpiece Y, Blondel B, Crainic R, van der Werf, Streeck RE (1986) A poliovirus neutralization epitope expressed on hybrid hepatitis B surface antigen particles. Science 233: 472–475
Delpeyroux F, Chenciner N, Lim A, Lambert M, Malpiece Y, Streeck RE (1987) Insertions in the hepatitis B surface antigen: effect on assembly and secretion of 22-nm particles from mammalian cells. J Mol Biol 195: 343–350
Delpeyroux F, Peillon N, Blondel B, Crainic R, Streeck RE (1988) Presentation and immunogenicity of the hepatitis B surface antigen and poliovirus neutralization antigen on mixed empty envelope particles. J Virol 62: 1836–1839
Dubois MF, Pourcel C, Rousset S, Chany C, Tiollais P (1980) Excretion of hepatitis B surface antigen particles from mouse cells transformed with cloned viral DNA. Proc Natl Acad Sci USA 77: 4549–4553
Eble BE, Lingappa VR, Ganem D (1986) Hepatitis B surface antigen: an unusual secreted protein initially synthesized as a transmembrane polypeptide. Mol Cell Biol 6: 1454–1463

Eble BE, Macrae DR, Lingappa VR, Ganem D (1987) Multiple topogenic sequences determine the transmembrane orientation hepatitis B surface antigen. Mol Cell Biol 7: 3591–3601
Eble BE, Lingappa VR, Ganem D (1990) The N-terminal (preS2) domain of a hepatitis B virus surface glycoprotein is translocated across membranes by downstream signal sequences. J Virol 64: 1414–1419
Enders GH, Ganem D, Varmus HE (1985) Mapping the major transcripts of ground squirrel hepatitis virus: the presumptive template for reverse transcriptase is terminally redundant. Cell 42: 297–308
Enders GH, Ganem D, Varmus HE (1987) 5′-terminal sequences influence the segregation of ground squirrel hepatitis virus RNAs into polyribosomes and viral core particles. J Virol 61: 35–41
Feitelson M, Marion P, Robinson W (1982) Core particles of hepatitis B virus and ground squirrel hepatitis virus II: characterization of the protein kinase reaction associated with ground squirrel hepatitis virus and hepatitis B virus. J Virol 43: 741–748
Galibert F, Mandart E, Fitoussi F, Tiollais P, Charnay P (1979) Nucleotide sequence of hepatitis B virus genome (subtype ayw) cloned in *E. coli*. Nature 281: 646–650
Galibert F, Chen TN, Mandart E (1982) Nucleotide sequence of a cloned woodchuck hepatitis virus genome: comparison with the hepatitis B virus sequence. J Virol 41: 51–65
Gallina A, Bonelli F, Zentilin L, Rindi G, Muttini M, Milanesi G (1989) A recombinant hepatitis B core antigen polypeptide with the protamine-like domain deleted self-assembles into capsid particles but fails to bind nucleic acids. J Virol 63: 4645–4652
Ganem D (1982) Persistent infection of humans with hepatitis B virus: mechanisms and consequences. Rev Infect Dis 4: 1026–1047
Ganem D, Varmus HE (1987) The molecular biology of the hepatitis B viruses. Annu Rev Biochem 56: 651–694
Garcia P, Ou J, Rutter WJ, Walter P (1988) Targeting of protein of hepatitis B virus to the endoplasmic reticulum membrane: after signal peptide cleavage translocation can be aborted and the product released into the cytoplasm. J Cell Biol 106: 1093–1104
Gavilanes F, Gonzales-Ros A, Peterson D (1982) Structure of hepatitis B surface antigen: characterization of the lipid components and their association with the viral proteins. J Biol Chem 257: 7770–7777
Gerber M, Hadziyannis S, Vissoulis C, Schaffner F, Paronetto F, Popper H (1974) Electron microscopy and immunoelectron microscopy of cytoplasmic hepatitis B antigen in hepatocyte. Am J Pathol 75: 489–502
Gerin J, Purcell R, Hoggan M, Holland P, Chanock R (1969) Biophysical properties of Australian antigen. J Virol 36: 787–795
Gerlich W, Goldmann U, Muller R, Stibbe W, Wolff W (1982) Specificity and localization of the hepatitis B virus-associated protein kinase. J Virol 42: 761–766
Gocke D (1975) Extrahepatic manifestation of viral hepatitis. Amer J Med Sci 207: 49–55
Heermann KH, Goldmann I, Schwartz W, Seyffarth T, Baumgarten H, Gerlich WH (1984) Large surface proteins of hepatitis B virus containing the preS sequence. J Virol 52: 396–402
Hemmingsen SM, Woolford C, van der Vies SM, Tilly K, Dennis DT, Georgopoulos CP, Hendrix RW, Ellis RJ (1988) Homologous plant and bacterial proteins chaperone oligomeric protein assembly. Nature 333: 330–334
Hirsch R, Lavine J, Chang L, Varmus H, Ganem D (1990) Polymerase gene products of hepatitis B viruses are required for genomic RNA packaging as well as for reverse transcription. Nature 344: 522–555
Hoofnagle JH, Gerety RJ, Barker LF (1975) Antibody to hepatitis B core antigen. Am J Med Sci 270: 179–187
Hoofnagle JH, Michalak T, Nowoslawski A, Gerety RJ, Barker (1978) Immunofluorescence microscopy in experimentally induced type B hepatitis in the chimpanzee. Gastroenterology 74: 182–187
Jean-Jean O, Levrero M, Will H, Perricaudet M, Rossignol JM (1989a) Expression mechanism of the hepatitis B virus (HBV) C gene and biosynthesis of HBe antigen. Virology 170: 99–106
Jean-Jean O, Salhi S, Carlier D, Elie C, DeRecondo A-M, Rossignol J-M (1989b) Biosynthesis of hepatitis B virus e antigen: directed mutagenesis of the putative aspartyl protease site. J Virol 63: 5497–5500
Junker-Niepmann M, Bartenschlager R and Schaller H (1990) A short *cis*-acting sequence is required for hepatitis B virus pregenome encapsidation and sufficient for packaging of foreign RNA EMBO 9: 3389–3396
Kamimura T, Yoshikawa A, Ichida F, Sasaki H (1981) Electron microscopy studies of Dane particles in

hepatocytes with special references to intracellular development of Dane particles and their relation with the HBeAg in the serum. Hepatology 1: 392–397
Kelly R (1985) Pathways of protein secretion in eukaryotes. Science 230: 25–31
Kodama K, Ogasawara N, Yoshikawa H, Murakami S (1985) Nucleotide sequence of a cloned woodchuck hepatitis virus genome: evolutional relationship between hepadnaviruses. J Virol 56: 978–986
Kozak M (1989) The scanning model for translation: an update. J Cell Biol 108: 229–241
Kuroki K, Russnack R, Ganem D (1989) A novel N-terminal amino acid sequence required for retention of a hepatitis B virus glycoprotein in the endoplasmic reticulum. Mole Cell Biol 9: 4459–4466
Kuroki K, Floreani M, Mimms L, Ganem D (1990) Epitope mapping of the preS1 domain of the hepatitis B virus large surface antigen. Virology 176: 620–624
Lanford R, Luckow V, Kennedy R, Dreesman G, Notvall L, Summers MD (1989) Expression and characterization of hepatitis B virus surface antigen polypeptides in insect cells with a baculovirus expression system. J Virol 63: 1549–1557
Laub O, Rall LB, Truett M, Shaul Y, Standring DN, Valenzuela P, Rutter WJ (1983) Synthesis of hepatitis B surface antigen in mammalian cells: expression of the entire gene and the coding region. J Virol 48: 271–280
Lavine J, Hirsch R, Ganem D (1989) A system for studying the selective encapsidation of hepadnaviral RNA. J Virol 63: 4257–4263
Lien JM, Aldrich CE, Mason WS (1986) Evidence that a capped oligoribonucleotide is the primer for duck hepatitis B virus plus-strand DNA synthesis. J Virol 57: 229–236
Liu CC, Yansura D, Levinson A (1982) Direct expression of hepatitis B surface antigen in monkey cells from an SV40 vector. DNA1: 213–221
Machida A, Kishimoto S, Ohumura H, Miyamoto H, Baba K, Oda K, Nakamura T, Miyakawa Y (1983) A hepatitis B surface antigen poly-peptide (P31) with the receptor for polymerized human as well as chimpanzee albumins. Gastroenterology 85:268–274
Machida A, Kishimoto S, Ohnuma H, Baba K, Ito Y, Miyamoto H, Funatsu G, Oda K (1984) A polypeptide containing 55 amino acid residues coded by the preS region of hepatitis B virus deoxyribonucleic bears the receptor for polymerized human as well as chimpanzee albumins. Gastroenterology 86: 910–918
MacRae D, Bruss V and Ganem D (1991) Myristylation of a duck hepatitis B virus envelope is essential for infectivity but not for virus assembly. Virology (in press)
Mason W, Halpern M, England J, Seal G, Egan J, Coates L, Aldrich C, Summers J (1983) Experimental transmission of duck hepatitis B virus. Virology 131: 375–384
Matsuda K, Satoh S, Ohori H (1988) DNA binding activity of hepatitis B e antigen polypeptide lacking the protamine-like sequence of nucleocapsid protein of human hepatitis B virus. J Virol 62: 3517–3521
McAleer WJ, Buynak EB, Maigetter RZ, Wampler DE, Miller WJ, Hilleman MR (1984) Human hepatitis B vaccine from recombinant genes. Nature 307: 178–180
McLachlan A, Milich DR, Raney AK, Riggs MG, Hughes JL, Sorge J, Chisari FV (1987) Expression of hepatitis B virus surface and core antigens: influences of preS and precore sequences. J Virol 61: 683–692
Meric C, Spahr P (1986) Rous sarcoma virus nucleic acid binding protein p12 is necessary for viral 70 S RNA dimer formation and packaging. J Virol 60: 450–459
Michel M-L, Pontisso P, Sobczak E, Malpiece Y, Streeck RE, Tiollais P (1984) Synthesis in animal cells of hepatitis B surface antigen particles carrying a receptor for polymerized human serum albumin. Proc Natl Acad Sci USA 81: 7708–7712
Michel M-L, Mancini M, Sobczak E, Favier V, Guetard D, Bahraoui EM, Tiollais P (1988) Induction of antihuman immunodeficiency virus (HIV) neutralizing antibodies in rabbits immunized with recombinant HIV-hepatitis B surface antigen particles. Proc Natl Acad Sci USA 85: 7957–7961
Milich DR, Thornton GB, Neurath AR, Kent SB, Tiollais MMP, Chisari FV (1985) Enhanced immunogenicity of the preS region of hepatitis B surface antigen. Science 228: 1195–1198
Miller RH (1987) Proteolytic self-cleavage of hepatitis B virus core protein may generate serum E antigen. Science 236: 722–724
Miyanohara A, Imamura T, Araki M, Sugawara K, Ohtomo N, Matsubara K (1986) Expression of hepatitis B virus core antigen gene in Saccharomyces cerevisiae: synthesis of two polypeptides translated from different initiation codons. J Virol 59: 176–180
Moriarty AM, Hoyer BH, Shih JWK, Gerin JL, Hamer DH (1981) Expression of the hepatitis B virus

surface antigen gene in cell culture by using a simian virus 40 vector. Proc Natl Acad Sci USA 78: 2606–2610
Nassal M, Galle PR, Schaller H (1989) Proteaselike sequence in hepatitis B virus core antigen is not required for e antigen generation and may not be part of an aspartic acid-type protease. J Virol 63: 2598–2604
Nassal M, Junker-Niepmann M and Schaller H (1990) Translational inactivation of RNA function: Discrimination against a subset of genomic transcripts during HBV nucleocapsid assembly. Cell 63: 1357–1363
Okada K, Kamiyama I, Inomata IM, Imai M, Miyakawa Y, Mayumi M (1976) E-antigen and anti-E in the serum of asymptomatic carrier mothers as indicators of positive and negative transmission of hepatitis B virus to their infants. N Engl J Med 294: 746–749
Okamoto H, Yotsumoto S, Akahane Y, Yamanaka T, Miyakana Y and Mayumi M (1990) Hepatitis B viruses with precore region defects prevail in persistently infected hosts along with seroconversion to antibody against e antigen. J Virol 64: 1298–1303
Ou J-H, Rutter WJ (1987) Regulation of secretion of the hepatitis B virus major surface antigen by the preS1 protein. J Virol 61: 782–786
Ou J, Laub O, Rutter WJ (1986) Hepatitis B gene function: the precore region targets. Proc Natl Acad Sci USA 83: 1578–1582
Ou J, Yeh C, Yen TSB (1989) Transport of hepatitis B virus precore protein into the nucleus after cleavage of its signal peptide. J Virol 63: 5238–5243
Pasek M, Goto T Gilbert W, Zink B, Schaller H, Mackay P, Ledbetter G, Murray K (1978) Hepatitis B virus genes and their expression in *E. coli.* Nature 282: 575–579
Patzer EJ, Nakamura GR, Yaffe A (1984) Intracellular transport and secretion of hepatitis B surface antigen in mammalian cells. J Virol 51: 346–353
Patzer EJ, Nakamura GR, Simonsen GC, Levinson AD, Brands R (1986) Intracellular assembly and packaging of hepatitis B surface antigen particles occur in the endoplasmic reticulum. J Virol 58: 884–892
Perara E, Rothman RE, Lingappa VR (1986) Uncoupling translocation from translation implications for transport of proteins across membranes. Science 232: 348–352
Persing DH, Varmus HE, Ganem D (1985) A frameshift mutation in the preS region of the human hepatitis B virus genome allows production of surface antigen particles but eliminates binding to polymerized albumin. Proc Natl Acad Sci USA 82: 3440–3444
Persing DH, Varmus HE, Ganem D (1986) Antibodies to preS and X determinants arise during natural infection with ground squirrel hepatitis virus. J Virol 60: 177–184
Persing DH, Varmus HE Ganem D (1987) The preS1 protein of hepatitis B virus is acylated at its amino terminus with myristic acid. J Virol 61: 1672–1677
Peterson DL (1981) Isolation and characterization of the major protein and glycoprotein of hepatitis B surface antigen. J Biol Chem 256: 6975–6983
Petit MA, Pillot J (1985) HBc and HBe antigenicity and DNA-binding activity of major core protein P22 in hepatitis B virus core particles isolated from the cytoplasm of human liver cells. J Virol 53: 543–551
Pugh JC, Sninsky JJ, Summers JW, Schaeffer E (1987) Characterization of a preS polypeptide on the surfaces of infectious avian hepadnavirus particles. J Virol 61: 1384–1390
Pugh JC, Zweidler A, Summers JW (1980) Characterization of the major duck hepatitis B virus core particle protein. J Virol 63: 1371–1376
Robinson WS, Greenman R (1974) DNA polymerase in the core of the human hepatitis B virus candidate. J Virol 13: 1231–1236
Robinson WS, Lutwick LI (1976) The virus of hepatitis type B. N Engl J Med 295: 1168–1175
Roossinck MJ, Siddiqui A (1987) In vivo phosphorylation and protein analysis of hepatitis B virus core antigen. J Virol 61: 955–961
Roossinck MJ, Jameel S, Loukin SH, Siddiqui A (1986) Expression of hepatitis B viral core region in mammalian cells. Mol Cell Biol 6: 1393–1400
Schlicht HJ, Schaller H (1989) The secretory core protein of human hepatitis B virus is expressed on the cell surface. J Virol 63: 5399–5404
Schlicht HJ, Kuhn C, Guhr B, Mattaliano RJ, Schaller H (1987a) Biochemical and immunological characterization of the duck hepatitis B virus envelope proteins. J Virol 61: 2280–2285
Schlicht HJ, Salfeld J, Schaller H (1987b) The duck hepatitis B virus preC region encodes a signal sequence which is essential for synthesis and secretion of processed core proteins but not for virus formation. J Virol 61: 3701–3709

Schlicht HJ, Bartenschlager R, Schaller H (1989) The duck hepatitis B virus core protein contains a highly phosphorylated C-terminus that is essential for replication but not for RNA packaging. J Virol 63: 2995–3000

Seeger C, Maragos J (1990) Identification and characterization of the woodchuck hepatitis virus origin of DNA replication. J Virol 64: 16–23

Seeger C, Ganem D, Varmus HE (1984) Nucleotide sequence of an infectious molecularly cloned genome of ground squirrel hepatitis virus. J Virol 51: 367–375

Seeger C, Ganem D, Varmus HE (1986) Genetic and biochemical evidence for the hepatitis B virus replication strategy. Science 232: 477–485

Sells MA, Chen ML, Acs G (1987) production of hepatitis B virus particles in Hep G2 cells transfected with cloned hepatitis B virus DNA. Proc Natl Acad Sci USA 84: 1005–1009

Shaeffer E, Snyder R, Sninsky J (1986) Identification and localization of preS-encoded polypeptides from woodchuck and ground squirrel hepatitis viruses. J Virol 57: 173–182

Shih J, Gerin J (1977) Proteins of hepatitis B surface antigen: amino acid compositions of the major polypeptides. J Virol 21: 1219–1222

Simon K, Lingappa V, Ganem D (1988a) Secreted hepatitis B surface antigen polypeptides are derived from a transmembrane precursor. J Cell Biol 107: 2163–2168

Simon K, Lingappa V, Ganem D (1988b) A block to the intracellular transport and assembly of hepatitis B surface antigen polypeptides in Xenopus oocytes. Virology 166: 76–81

Speiss M, Lodish H (1986) An internal signal sequence: the asialoglycoprotein membrane anchor. Cell 44: 177–185

Stah SJ, Murray K (1989) Immunogenicity of peptide fusions to hepatitis B virus core antigen. Proc Natl Acad Sci USA 86: 6283–6287

Standring DN, Rutter WJ, Varmus HE, Ganem D (1984) Transcription of the hepatitis B surface antigen gene in cultured murine cells initiates within the presurface region. J Virol 50: 563–571

Standring DN, Ou J, Rutter WJ (1986) Assembly of viral particles in Xenopus oocytes: pre-surface-antigens regulate secretion of the hepatitis B viral surface envelope particle. Proc Natl Acad Sci USA 83: 9338–9342

Standring DN, Ou J, Masiarz F, Rutter WJ (1988) A signal peptide encoded within the precore region of hepatitis B virus directs the secretion of a heterogeneous population of e antigens in Xenopus oocytes. Proc Natl Acad Sci USA 85: 8405–8409

Stenlund A, Lamy D, Moreno-Lopez J, Ahola H, Pettersson U, Tiollais P (1983) Secretion of the hepatitis B virus surface antigen from mouse cells using an extra-chromosomal eukaryotic vector. EMBO J 2: 669–673

Stibbe W, Gerlich W (1983) Structural relationships between minor and major proteins of hepatitis B surface antigen. J Virol 46: 626–628

Stratowa C, Doehmer J, Wang Y, Hofschneider PH (1982) Recombinant retroviral DNA yielding high expression of hepatitis B surface antigen. EMBO J 1: 1573–1578

Summers J, Mason WS (1982) Replication of the genome of a hepatitis B-like virus by reverse transcription of an RNA intermediate. Cell 29: 403–415

Summers J, O'Connell A, Millman I (1975) Genome of hepatitis B virus: restriction enzyme cleavage and structure of DNA extracted from Dane particles. Proc Natl Acad Sci USA 72: 4597–4601

Sureau C, Romet-Lemonne JL, Mullins JI, Essex M (1986) Production of hepatitis B virus by a differentiated human hepatoma cell line after transfection with cloned circular HBV DNA. Cell 47: 37–47

Szmuness W, Stevens CE, Harley EJ, Zang EA, Oleszko WR, William DC, Sadovsky R, Morrison JM (1980) Hepatitis B vaccine. N. Engl J Med 303: 833–841

Towler D, Gordon J, Adams S, Glaser L (1988) The biology and enzymology of eukaryotic protein acylation. Annu Rev Biochem 57: 69–99

Townsend A, Bodmer H (1989) Antigen recognition by class I-restricted T lymphocytes. Annu Rev Immunol 7: 601–624

Tsurimoto T, Fujiyama A, Matsubara K (1987) Stable expression and replication of hepatitis B virus genome in an integrated state in a human hepatoma cell line transfected with the cloned viral DNA. Proc Natl Acad Sci USA 84: 444–448

Tuttleman J, Pourcel C, Summers J (1986) Formation of the pool of covalently closed circular viral DNA in hepadnavirus-infected cells. Cell 47: 451–460

Uy A, Bruss V, Gerlich W, Kochel H, Thomson R (1986) Precore sequence of hepatitis B virus including a antigen and membrane association of the viral core protein. Virology 155: 89–96

Valenzuela P, Gray P, Quiroga M, Zaldivar J, Goodman HM, Rutter WJ (1979) Nucleotide sequence of the gene coding for the major protein of hepatitis B virus surface antigen. Nature 280: 815–819

Valenzuela P, Quiroga M, Zaldivar J, Gray P, Rutter WJ (1980) The nucleotide sequence of the hepatitis B viral genome and the identification of the major viral genes. In: Fields BN, Jaenisch R (eds) Animal virus genetics ICN/UCLA Symp Mol Cell Biol, pp 57–70
Valenzuela P, Medina A, Rutter WJ (1982) Synthesis and assembly of hepatitis B virus surface antigen particles in yeast. Nature 298: 347–351
Valenzuela P, Coit D, Medina-Selby A, Kuo C, Van Nest G, Burke RL, Bull P, Urdea M, Graves P (1985) Antigen engineering in yeast: Synthesis and assembly of hybrid hepatitis B surface antigen—herpes simply 1 gD particles. Biotechnology 3: 323–326
Wampler DE, Lehman ED, Boger J, McAleer WJ, Scolnick EM (1985) Multiple chemical forms of hepatitis B surface antigen produced in yeast. Proc Natl Acad Sci USA 82: 6830–6834
Weimer T, Salfeld J, Will H (1987) Expression of hepatitis B virus core gene in vitro and in vivo. J Virol 61: 3109–3113
Will H, Cattaneo R, Pfaff E, Kuhn C, Roggendorf M, Schaller H (1984) Expression of hepatitis B antigen with a simian virus 40 infection. J Virol 50: 335–342
Will H, Reiser W, Weimer T, Pfaff E, Buscher M, Sprengel R, Cattaneo R, Schaller H (1987) Replication strategy of human hepatitis B virus. J Virol 61: 904–911
Wong DT, Nath N, Sninsky JJ (1985) Identification of hepatitis B virus polypeptides encoded by the enitre preS open reading frame. J Virol 55: 223–231
Yaginuma K, Shirakata Y, Kobayashi M, Koike K (1987) Hepatitis B virus (HBV) particles are produced in a cell culture system by transient expression of transfected BHV DNA. Proc Natl Acad Sci USA 84: 2678–2682
Yamada G, Feinberg L, Nakane P (1978) Hepatitis B: cytologic localization of virus antigens and the role of the immune response. Human Path 9: 93–109
Yeh C-T, Liaw Y-F and Ou J-H (1990) The arginine-rich domain of HBV precore and core proteins contains a signal for nuclear transport. J virol 64: 6141–6147

Analysis of Hepadnavirus Gene Expression, Biology, and Pathogenesis in the Transgenic Mouse

F. V. CHISARI

1 Introduction

Hepatitis B virus (HBV) causes acute and chronic hepatocellular injury and hepatocellular carcinoma (reviewed in CHISARI et al. 1989b). During acute infection the virus replicates episomally within infected cells. In chronic infection the virus may also integrate into the host cell genome. In HBV-induced hepatocellular carcinoma, integrated viral DNA sequences are virtually always present, and the viral and flanking cellular sequences may be rearranged (see chapter by C.E. ROGLER, this volume). Many aspects of the biology of HBV infection and the mechanisms whereby it causes liver cell injury and malignant transformation are not well understood, in large part because of the absence of appropriate tissue culture and animal models to study stages of this process. With the recent advent of transgenic mouse technology and the ability to introduce parts or the entirety of the HBV genome into the mouse germline, many of these questions are now approachable.

Research Institute of Scripps Clinic, 10666 North Torrey Pines Road, La Jolla, CA 92037, USA

Current Topics in Microbiology and Immunology, Vol. 168

During the past few years, several laboratories have taken advantage of this opportunity to expand our understanding of the biology of this virus and to elucidate some of the reciprocal interactions of virus and host that may determine the outcome of infection. In this chapter I will review this new information and discuss its implications in terms of the biology and pathogenesis of HBV. Since the model system entails the introduction of the HBV genome or its constituent open reading frames (ORFs) into transgenic animals, a brief overview of selected aspects of the viral genome will be presented before describing the studies themselves. This background material is reviewed in greater detail in other chapters of this volume.

2 Products of the HBV Genome

The HBV virion contains a circular, partially double-stranded DNA genome approximately 3200 base pairs in length (Fig. 1). The long (minus) strand encodes the viral mRNA and the viral pregenomic RNA which is reverse transcribed as an early step in viral replication. The minus strand contains four long, overlapping ORFs that encode structural (envelope, nucleocapsid) and nonstructural (polymerase, transactivating protein) viral elements.

The nucleocapsid ORF contains two in-phase start codons that define two overlapping polypeptides, the shorter of which (core, HBcAg) is a cytoplasmic and nuclear protein which represents the viral nucleocapsid. Recent studies indicate that HBcAg is expressed on the surface of circulating infectious virus particles (MOLLER et al. 1989). Its longer counterpart (precore) is a secreted, cytoplasmic, nuclear, and membrane protein (OU et al. 1989; SCHLICHT and SCHALLER 1989), whose function is currently undefined. Based on considerable circumstantial clinical evidence, it is believed that a cellular immune response to the nucleocapsid antigens is an important factor in viral clearance in HBV infection (reviewed in CHISARI et al. 1989b).

The envelope ORF contains three in-phase translation start codons which define the amino termini of three overlapping polypeptides, whose expression is transcriptionally regulated (see chapter by H. SCHALLER, this volume). All three are components of the virus envelope. In addition, the major and middle envelope polypeptides assemble into small, 22-nm spherical particles by budding into the lumen of the endoplasmic reticulum (ER) and they are rapidly secreted after passage and glycosylation in the Golgi apparatus (PATZER et al. 1984, 1986; EBLE et al. 1986, 1987). The large envelope polypeptide (L) also exerts an important structural influence on the formation of the abundant, noninfectious, subviral particles that are characteristic of HBV infection (CHISARI et al. 1986; STANDRING et al. 1986; MCLACHLAN et al. 1987; OU and RUTTER 1987; MOLNAR-KIMBER et al. 1988). Depending on the relative molar ratio of the large envelope polypeptide to the major and middle envelope polypeptides within a

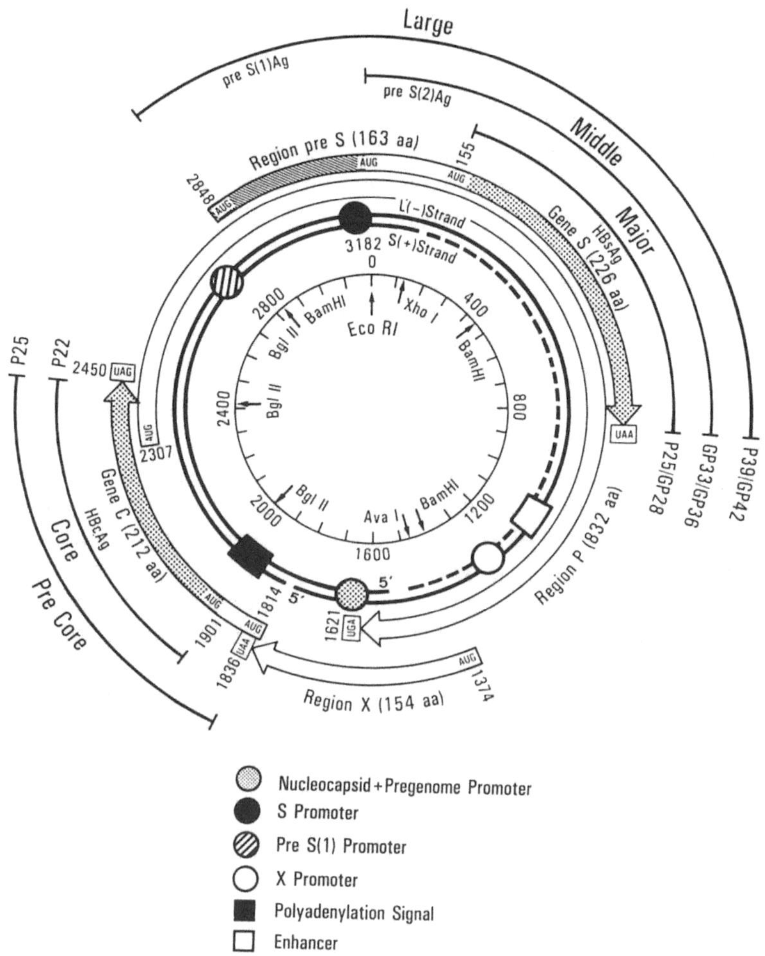

Fig. 1. Map of hepatitis B virus genome

given cell, either short secretable filaments or long, nonsecretable, branching filaments that accumulate within the ER are produced. Intracellular accumulation of these nonsecretable filaments is cytopathic, producing so-called ground glass hepatocytes when this process occurs in infected liver cells. In fact, this process may ultimately culminate in the death of the cell (CHISARI et al. 1987). Recent evidence (MORIYAMA et al. 1990) indicates that the envelope-specific, cytotoxic T cells can also cause hepatocellular injury, at least in transgenic mice, suggesting that these envelope polypeptides may also be important targets for immune-mediated liver disease during HBV infection in humans.

The X ORF encodes a transcriptional *trans*-activating protein that positively regulates transcription from HBV and other viral and cellular promoters (TWU and SCHLOEMER 1987; SPANDAU and LEE 1988; SETO et al. 1988; TWU et al. 1989;

COLGROVE et al. 1989). Current data suggest that the X-gene product can transactivate a broad assortment of promoters in many cell types and that it may operate in a fashion reminiscent of the adenovirus E1A transactivator, by influencing the transcription process in a general fashion rather than by directly binding target sequences (COLGROVE et al. 1989). Because of these properties it has been proposed that dysregulated expression of the X-gene product may be involved in hepatocarcinogenesis in chronic HBV infection (WOLLERSHEIM et al. 1988).

Four promoter elements regulate expression of the various HBV genes (chapter by H. SCHALLER, this volume). The core promoter regulates synthesis of the RNA pregenome, the template for reverse transcription (chapter by C. SEEGER et al. this volume), as well as a slightly larger RNA(s). Together, these RNAs encode the precore, core, and polymerase polypeptides. Two independent promoters control envelope gene expression. The *preS promoter* is relatively silent transcriptionally in comparison with the core and S promoters (BUSCHER et al. 1985; OU and RUTTER 1985; SIDDIQUI et al. 1986; WILL et al. 1987). It is located just upstream of the first translational start codon in the envelope region, and it encodes a 2.4-kb transcript from which the large envelope polypeptide is produced. The *S promoter* is located within the envelope coding region just upstream of the translation start site of the middle envelope polypeptide (RALL et al. 1983; ENDERS et al. 1985, 1987). It is transcriptionally very active and encodes a 2.1-kb transcript that, by virtue of microheterogeneity at its 5′ end, controls production of the middle and major envelope polypeptides. The usual balance between transcription initiation at the S and preS promoters normally leads to the relative underproduction of the large envelope polypeptide with respect to the two smaller proteins. The low level of preS promoter activity may be due to a downstream, negative regulatory element (BULLA and SIDDIQUI 1988). It is now clear that dysregulated overproduction of the large envelope mRNA, with concomitant overproduction of the large envelope polypeptide, not only inhibits assembly and secretion of viral particles but, as mentioned above, has significant pathological consequences. Like the preS promoter, the *X gene promoter* is also usually transcriptionally quite silent in vivo (KANEKO and MILLER 1988). In view of the transcriptional transactivating properties of the X-gene product, dysregulated X-gene expression may also have important pathophysiological consequences, including hepatocarcinogenesis.

3 A Model of HBV-Associated Liver Disease and Hepatocellular Carcinoma

The pathogenetic mechanisms responsible for liver injury following HBV infection are not well understood. Historically, the virus was thought not to be directly cytopathic; however, as already mentioned, studies of L protein overexpression suggest that this is not strictly true. It has also been suggested that the balance between the spread of infection through the liver and the quality and vigor of the antiviral immune response may determine the severity and the duration of HBV-induced liver injury. The mechanism(s) responsible for malignant transformation is (are) also incompletely defined. Since the interval between infection and hepatoma is usually several decades, it appears that HBV is neither directly nor acutely oncogenic.

With the advent of embryo microinjection technology it became evident that many of these questions relating to HBV pathogenesis and hepatocellular transformation might be directly examined by introduction of partial or complete copies of the HBV genome into transgenic mice. Thus, several groups of investigators have independently generated lineages of transgenic mice containing HBV sequences, and these studies will now be reviewed.

3.1 Microinjected Constructs

Several different constructs derived from the HBV genome have been used to produce transgenic mice. All of the constructs published thus far (BABINET et al. 1985; CHISARI et al. 1985; YAMAMURA et al. 1987; BURK et al. 1988; FARZA et al. 1988; ARAKI et al. 1989b; HINO et al. 1989) have contained at least the envelope and X transcription units and the intervening enhancer element (see Fig. 1). In two instances (CHISARI et al. 1986, 1987), the preS1 promoter was deleted from this subgenomic fragment and replaced by the mouse metallothionein or the mouse albumin promoter elements, which permit deregulated expression of the large envelope polypeptide (see below). In other experiments (FARZA et al. 1988; ARAKI et al. 1989b), greater than genome length constructs designed to be compatible with expression of all viral transcription units were used. Many laboratories are currently attempting to produce transgenic mice by injection of multiple subgenomic fragments capable of encoding separately the nucleocapsid, X, and polymerase proteins of HBV, but progress in these areas is too preliminary to warrant discussion at this time.

3.2 Tissue-Specific Expression in Transgenic Mice

Using constructs containing only HBV-derived regulatory sequences, the several lineages of transgenic mice that have been produced (BABINET et al. 1985; BURK et al. 1988; FARZA et al. 1988; ARAKI et al. 1989b) demonstrate preferential expression of HBV-encoded transcripts in the liver (hepatocytes) and kidney (proximal convoluted tubules). It is notable that in most lineages, expression is not limited to the liver, that it is often greater in the kidney than in the liver, and that in multiple lineages it extends to other tissues including the heart, stomach, small intestine, pancreas, brain, and testis, albeit at lower levels. Curiously, HBV expression has not been detected in the thymus, spleen, or bone marrow in any of the existing transgenic lineages despite evidence of viral replication and expression in lymphoid tissues in infected humans and woodchucks (KORBA et al. 1989). Other than obvious species differences, the reason for this inconsistency is not known.

The fairly widespread expression of the transgenic HBV genome in many tissues together with mounting evidence of extrahepatic hepadnavirus expression during natural infection (KORBA et al. 1986) suggests that the relative tissue and species specificity of HBV is a reflection of multiple constraints at the levels of viral entry, replication, and gene expression, and that none of these constraints is absolutely specific for the human hepatocyte. It also appears that the HBV genome contains *cis*-acting regulatory elements that are preferentially responsive to tissue-specific transacting factors that are particularly abundant in the liver and kidney.

3.3 Viral Replication in Transgenic Mice

Following microinjection of greater-than-genome length constructs capable of encoding all of the viral transcripts as well as the pregenomic viral RNA, several groups have demonstrated the presence of small amounts of cytoplasmic single-stranded and double-stranded HBV DNA replicative intermediates in liver, kidney, and heart and complete virus-like particles in the serum (FARZA et al. 1988; ARAKI et al. 1989b; CHISARI et al., unpublished data), indicative of low level viral replication. During natural infection, the cytoplasmic replicative intermediates are transported within nascent core particles to the nucleus where they mature into covalently closed circular (supercoiled) viral DNA molecules that serve to amplify the transcriptional template for viral replication (chapter by C. SEEGER et al., this volume). For unknown reasons, this process is inefficient in transgenic mice, and transcription is limited to the small number of integrated transgenes present in each cell. The absence of intranuclear, supercoiled, viral genomes may also contribute to the apparent absence of secondary viral integrations in these lineages.

The 2.1-kb (HBs) transcript is most abundant and widespread in its tissue distribution in these lineages. The 3.5-kb (RNA pregenome/nucleocapsid) transcript (chapter by H. SCHALLER and M. FISCHER, this volume) is usually less abundant and is preferentially expressed in liver, kidney, and heart. Viral DNA replication is, indeed, limited to these three tissues, a finding consistent with the fact that DNA synthesis initiates with reverse transcription of pregenomic RNA. As in natural infection, the 2.4-kb (preS) and the 0.8-kb (X) transcripts are rare in transgenic mice. The preS transcript has been found only in the liver and kidney where it presumably contributes to the formation of complete virus particles. The X transcript is also present in liver and kidney where it presumably augments the expression of the remaining viral gene products. For unknown reasons, the X transcript is also commonly found in the testis where it is usually the only HBV-encoded transcript present.

3.4 Regulation of HBV Gene Expression in Transgenic Mice

Integration of HBV sequences is a natural consequence of chronic infections and is generally believed to play some role in the final outcome of the disease. As discussed in the chapter by C.E. ROGLER (this volume), integration may have a direct role in carcinogenesis. Expression from integrated DNA also explains the continued synthesis of HBsAg when all episomal forms of HBV DNA have disappeared. It is less clear why HBcAg is not expressed, and there is still uncertainty about the expression of *pol* and X-gene products from integrated DNA. The possible expression of X from integrated DNA is especially intriguing because of speculation about the role of this transactivator in the development of liver cancer. Unfortunately, analysis of these issues in infected livers presents practical and technical problems that are virtually insurmountable. Some of them are approachable in the transgenic mouse.

Analysis of assorted independent transgenic lineages demonstrates that HBV gene expression is regulated at multiple levels. Several groups have documented an association between the degree of methylation and transcription. Indeed, lack of HBV expression has been associated with hypermethylation of the transgene sequences in numerous transgenic lineages (YAMAMURA et al. 1987; HADCHOUEL et al. 1987; FARZA et al. 1988; POURCEL 1989; EL-GHOR and BURK 1989; ARAKI et al. 1989a). Further, HBV sequences are often hypomethylated in tissues that express the corresponding HBV genes in contrast to those that do not; these associations, however, are not absolute (POURCEL 1989; EL-GHOR and BURK 1989). Demethylation of HBV sequences has been associated with the onset of HBV gene expression during embryonic development (POURCEL 1989). Similarly, demethylation in vivo by treatment with 5-azacytidine is associated with expression of HBsAg in transgenic mice that carry heavily methylated, nonexpressed HBV sequences (ARAKI et al. 1989a). Finally, a transgenic lineage has been described in which HBV expression is irreversibly extinguished when the integrated HBV sequences are methylated de novo upon passage through the female germline (HADCHOUEL et al. 1987). Although this observation is unique to a single transgenic lineage and is, therefore, most probably a position effect, it further underscores the potential relevance of hypermethylation to suppression of HBV gene expression during the course of natural HBV infections, e.g., in the regulation of acute infections or in the shut-off of virus production late in the course of chronic infections.

As in other systems, hypomethylation may be necessary, but it clearly is not sufficient for the expression of integrated HBV DNA. In transgenic mice and during chronic HBV infection, expression of integrated viral DNA sequences probably depends upon the cooperative effects of multiple additional factors such as chromatin structure, cellular flanking sequences, and exposure to cellular- and virus-encoded DNA-binding proteins that regulate viral gene expression in addition to the methylation status of the integrated DNA. Along these lines, EL-GHOR and BURK (1989) have recently demonstrated a positive correlation between tissue-specific expression of HBV sequences and the presence of DNase I-hypersensitive sites in the HBV enhancer element and the core promoter region in chromatin obtained from different tissues of two transgenic mouse lineages. If the presence and location of DNase 1-hypersensitive sites reflects a change in chromatin structure secondary to local transcriptional activity, it is possible that they represent the site of interaction between regulatory regions of the viral genome and cellular transcription factors. Since several cellular transcription factors have been described in recent years that interact with HBV sequences in the enhancer and core promoter regions (chapter by H. SCHALLER, this volume), it is reasonable to suspect that similar interactions may contribute to the expression of the integrated viral genome in these transgenic mice.

The influence of flanking cellular sequences on transgene expression (position effect) is dramatically illustrated by the integration site-dependent maternal imprinting of HBV gene expression in a transgenic lineage described above (HADCHOUEL et al. 1987). It is further shown in the variable frequency and distribution of HBsAg-positive hepatocytes in 10 independent transgenic lineages containing the identical transgene at different integration sites (CHISARI et al. 1987). The percentage of HBsAg-positive hepatocytes in these different transgenic lineages varies between 1% and 100%, and their distribution throughout the hepatic lobule may be either central, peripheral, focal, or universal in a lineage-dependent manner, and these features breed true from generation to generation. The only meaningful variable in these experiments that can be responsible for these differences in expression is the transgene integration site. These observations imply that HBV gene expression following integration during natural HBV infection is determined not only by the structure of the integrated sequences and the differentiation state of the cell but also by the genetic and structural characteristics of the integration site itself. They also suggest that part of the transcriptional heterogeneity of hepatocytes may be due to factors other than blood flow-induced hormonal and nutrient gradients.

HBV gene expression in transgenic mice is also developmentally regulated (FARZA et al. 1987; DELOIA et al. 1989), with HBV-specific transcripts being first detected on day 14–15 of embryogenesis and increasing gradually thereafter until birth, with a dramatic increase perinatally, perhaps due to the large amounts of glucocorticoids present at that time. Indeed, injection of dexamethasone leads to a dramatic increase in serum HBsAg levels in transgenic mice (CHISARI et al. 1986; FARZA et al. 1987). This observation may rationalize the detection of a glucocorticoid-responsive element in HBV by TUR-KASPA et al. (1986) in analyses of viral expression in established cell lines.

One of the characteristics of HBV infection is its relative prevalence (ca. fivefold) among males as compared with females. Several groups (BABINET et al. 1985; FARZA et al. 1987; DELOIA et al. 1989; CHISARI et al. 1989a) have shown that HBV envelope expression is greater in male than in female transgenic mice. Indeed, liver HBV mRNA and serum HBsAg levels rise dramatically in male transgenic mice at puberty, presumably in response to sex hormonal induction. Accordingly, it has been shown (FARZA et al. 1987; DELOIA et al. 1989) that castration causes a decline in expression that is reversible by administration of androgens and, to a lesser extent, by estrogens. Thus, HBV gene expression is hormonally regulated in vivo, and it is conceivable that the male prevalence of HBV infection in the natural setting is at least partially determined at this level. The data also raise the possibility that hormonal changes such as those associated with stress, inflammation, etc. might induce changes in hepatic HBV gene expression that, in turn, could influence aspects of viral replication and pathogenesis in a heretofore unsuspected manner (see below).

4 Liver Disease in HBV Transgenic Mice

As noted at the beginning of this section, the transgenic mouse model provides us with the opportunity of studying liver disease mediated directly by the expression of viral gene products, an issue which is almost impossible to study in natural HBV infections. Moreover, because these animals are inbred it is possible to study specific components of immune-mediated injury by transplantation into transgenic mice of well-characterized immune-effector cells from nontransgenic mice primed with various viral antigens. Thus far, my laboratory has concentrated on the HBV envelope region. We have used constructs in which a HBV *Bgl* II fragment (nt 2839-1986, Fig. 1) containing the entire HBV envelope coding region, the viral enhancer, and the HBV X region has been cloned downstream of relatively liver-specific regulatory elements that are either inducible (metallothionein, MT) or highly constitutively active (albumin, Alb) in the mouse (CHISARI et al. 1985, 1986, 1987, 1989a). Within these constructs the MT or Alb promoters take the place of the pre-S promoter that normally directs synthesis of the mRNA for the large envelope protein. One lineage that contains the metallothionein-HBV envelope (MT-env) construct and 10 lineages that construct and 10 lineages that contain the albumin-HBV envelope (Alb-env) construct have been produced, and selected lineages have been characterized in detail for evidence of progressive liver damage.

4.1 Role of the HBV Large Envelope Protein as a Mediator of Liver Disease and Hepatocellular Carcinoma

When the middle and major envelope polypeptides are produced in excess, they assemble into small (22-nm), spherical particles that are formed by internal budding into the lumen of the ER and are readily secreted (EBLE et al. 1986; SIMON et al. 1988). When produced at a roughly equimolar ratio with respect to the other envelope polypeptides, the large envelope polypeptide prevents the formation of small spherical HBsAg particles and causes the formation of long, branching HBsAg filaments (containing all 3 envelope polypeptides), which become trapped in the ER of the hepatocyte, thereby inhibiting HBsAg secretion from it (CHISARI et al. 1987). This negative effect on secretion has also been observed by other investigators in cell culture systems (STANDRING et al. 1986; OU and RUTTER 1987; MARQUARDT et al. 1987; MCLACHLAN et al. 1987; MOLNAR-KIMBER et al. 1988), but its biological significance was defined in the transgenic mouse model because it permitted expression in the hepatocyte, the normal target of HBV.

The most immediate consequence of overexpression of the large envelope protein, which occurs in some but not all of our transgenic lineages, is a dramatic expansion of the ER of the hepatocyte (CHISARI et al. 1987). This expansion, resulting from the accumulation of HBsAg filaments, eventually causes in transgenic mouse liver the ultrastructural and histologic changes characteristic of ground glass hepatocytes (GERBER et al. 1974) that are observed in the liver of patients with chronic HBV infection. The similarities between the human and the transgenic mouse ground glass cells are so strong, it is likely that they share a common molecular pathogenesis; in fact, the ER of ground glass cells in human liver contains the characteristic filamentous structures (GERBER et al. 1974). Studies with transgenic mice, described below, support the possibility that the ground glass cell of HBV infection may represent one point on a continuous spectrum of pathological changes ultimately resulting in liver tumors.

In particular, transgenic mice in lineages that overproduce the HBV large envelope polypeptide and accumulate toxic quantities of hepatitis B surface antigen within a large percentage of hepatocytes develop severe, prolonged, hepatocellular injury which initiates a programmed response within the liver, characterized by inflammation, regenerative hyperplasia, transcriptional deregulation, and aneuploidy, that inexorably progresses to neoplasia (CHISARI et al. 1989a; DUNSFORD et al. 1990). In view of the major structural changes in the ER, it is possible that the normal detoxification processes of hepatocytes, which are dependent upon proteins transported through this organelle, may be compromised; however, whether this or some other pathophysiological process is responsible for the death of the hepatocyte in these mice is currently unknown. The incidence of hepatocellular carcinoma observed in the different transgenic lineages corresponds to the frequency, severity, and age of onset of significant amounts of liver cell injury in each lineage, which themselves correspond to the intrahepatic concentration of HBsAg, a reflection of the number of hepatocytes expressing HBsAg, plus the level of HBsAg within each hepatocyte and are influenced by genetic background and sex. Thus, the inappropriate expression of a single viral structural gene product, large envelope protein, is sufficient to cause malignant transformation in this model.

In view of the prolonged antecedent injury and regeneration to replace dying hepatocytes, the early development of aneuploidy, and the altered transcriptional state (decreased HBsAg, increased alphafetoprotein, AFP, a marker of regeneration and transformation) of the regenerating and transformed hepatocytes as compared with healthy liver, it is likely that transformation occurs via a multistep process. Since transformation was observed in two independent lineages, without evidence of transgene rearrangement or instability, direct insertional activation of a cellular oncogene by the transgene is not likely. The link between injury and transformation is further strengthened by the fact that hepatocellular carcinoma has not been observed in any of the other published HBV transgenic lineages, none of which display evidence of liver injury (CHISARI et al. 1985; BABINET et al. 1985; FARZA et al. 1987; YAMAMURA et al. 1987; HADCHOUEL et al. 1987; BURK et al. 1988; FARZA et al. 1988; ARAKI et al. 1989b; HINO et al. 1989; DELOIA et al. 1989). In contrast, nodular regenerative activity has been described in livers of transgenic mice sustaining neonatal hepatitis as a consequence of the hepatocellular retention of alpha-1-antitrypsin (DYCAICO et al. 1988), further strengthening the link between injury and unregulated growth of hepatocytes.

These results suggest that severe, prolonged cellular injury, irrespective of etiology or pathogenesis, induces a preneoplastic proliferative response that fosters multiple random mutations or other chromosomal changes throughout

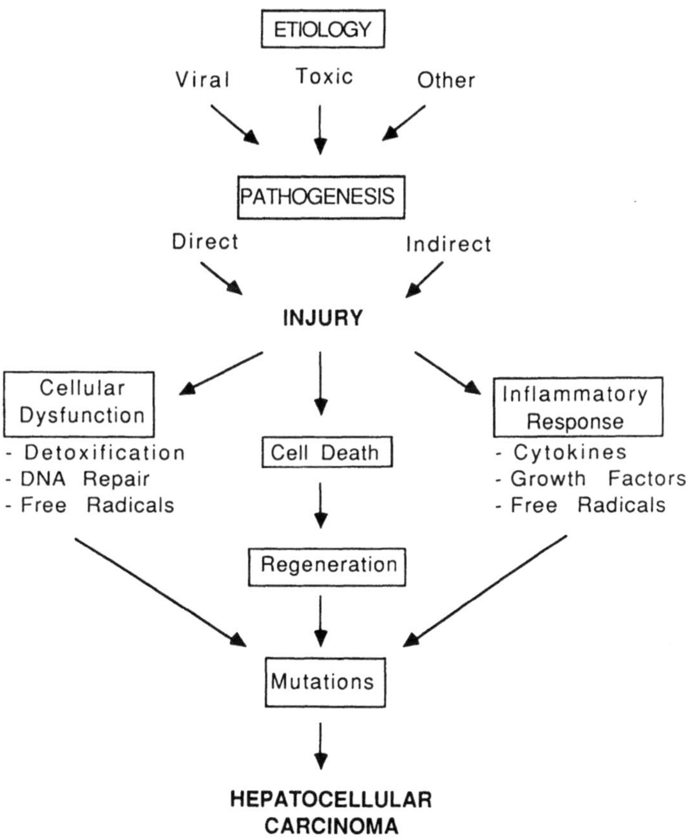

Fig. 2. Hypothetical mechanism of hepatocarcinogenesis

the cellular genome, which program the cell for unrestrained growth. Thus, our analyses of the transgenic mouse lead to the hypothesis that hepatocellular carcinoma in human HBV infection may often be due to secondary genetic events entirely unrelated to the integration of HBV DNA sequences, and that the acquisition of these secondary changes is directly related to the severity, extent, and duration of the virus-induced liver cell injury, whether due to immunological mechanisms or to the accumulation of directly cytotoxic viral gene products, or to other currently undefined events. In support of this hypothesis is the fact that human hepatocellular carcinoma occurs in the context of necrosis, inflammation, regeneration, and cirrhosis in several diseases other than hepatitis, such as chronic nonA-nonB hepatitis (ALTER 1988), alcoholism (LIEBER et al. 1986), hemochromatosis (NIEDERAU et al. 1985), glycogen storage disease (LIMMER et al. 1988), alpha-1-antitrypsin deficiency (CARLSON and ERIKSSON 1985; ERIKSSON et al. 1986), and primary biliary cirrhosis (MELIA et al. 1984). It is likely that transient hepatocellular regeneration alone is not sufficient to induce these putative genetic lesions since HCC does not develop following partial hepatectomy nor does it develop in the rapidly proliferating fetal liver. Rather, prolonged injury and regeneration seem to be of major importance. Thus, additional cofactors are probably needed for the development of HCC in this setting. Such cofactors might include mutagenic products of the associated inflammatory cells or impaired detoxification or DNA repair mechanisms within the injured hepatocyte as illustrated in Fig. 2.

4.2 Immunopathogenesis of Hepatocellular Injury

The HBV transgenic mouse model has created the opportunity to study not only liver cell injury mediated by expression of viral gene products but also to characterize the pathogenetic potential of the HBV-specific cytotoxic T cell (CTL) response. The role of CTLs in human hepatitis B has been inferred from the isolation from the liver (biopsies) and blood of helper and suppressor T cells reactive to specific viral antigens. It has not been possible, however, to demonstrate that these human lymphocytes actually play a role in liver injury. These issues can be studies in transgenic mice into which specific effector cells can be introduced by adoptive transfer (see below).

The minimal experimental model for T-cell-mediated liver cell injury requires hepatocellular synthesis of HBV gene products, the appropriate processing and association of these products with major histocompatibility determinants within the hepatocyte, the expression of the MHC-HBV peptide complex on the hepatocyte surface, the induction of a cellular immune response specific for one or more of the MHC-associated HBV peptides (i.e., the T-cell repertoire must include rearranged variable region genes specific for the expressed antigens, and it must be effectively induced), the appropriate functional T-cell subset must be expanded, and the target cell, i.e., the primary hepatocyte, must be susceptible to T-cell-mediated cytolysis. In view of the complexity of these requirements it is not surprising that the immunopathogenesis of HBV-induced liver disease is not yet defined.

Using the transgenic mouse model, we have demonstrated that HBV antigen-specific immune effector mechanisms can destroy HBV antigen-positive hepatocytes in vitro and, more importantly, in vivo (MORIYAMA et al. 1990). In these experiments, we used a transgenic mouse lineage (107-5) that expresses noncytotoxic amounts of the HBV large and major envelope polypeptides in the hepatocytes and, hence, does not develop spontaneous liver disease. (Lineage 107-5, like those that do overexpress L protein, contains a transgene in which the albumin promoter was placed in front of the HBV envelope.) This lineage develops liver cell injury, however, following the adoptive transfer of HBsAg-primed spleen cells or HBs-specific antibody from nontransgenic, syngeneic mice that have been immunized by infection with vaccinia-HBV envelope recombinants (MORIYAMA et al. 1990).

Injury mediated by HBs-specific antibody is rapid in onset and coincides with the disappearance of HBsAg and the appearance of anti-HBs in the serum. Clearly, this could be due either to complement-mediated cell lysis or to antibody-mediated cellular cytotoxicity (ADCC). A delayed form of liver cell injury occurs several days after the adoptive transfer of HBs antigen-primed spleen cells and is characterized histologically by widespread focal hepatocellular necrosis and a lymphomononuclear cell infiltrate with associated lymphoid infiltration in portal areas, thus resembling the histological picture of chronic hepatitis. In studies with fractionated spleen cell populations, we found that liver cell injury was mediated only by the T-cell population, specifically by the Lyt2 $(CD8)^+$ T-cell subset. A $CD8^+$, MHC class I-restricted, cytotoxic T-cell line that displayed HBV envelope-specific cytotoxicity was then established. The fine specificity of the CTL response was mapped to an epitope located between amino acids 21 and 40 of the major envelope polypeptide (Fig. 1). The HBsAg-specific CTL line caused liver cell injury when transferred into the transgenic mice in vivo. Injury was detectable as early as 5 days post transfer, and it persisted for several weeks thereafter, decreasing as a function of time. The data indicate that HBsAg-specific, MHC class I-restricted, $CD8^+$ CTL are functional in vivo and that endogenously synthesized HBV envelope antigens are expressed at the cell surface in a form that is recognized by such CTL.

Thus, studies with transgenic mice establish the scientific basis for the hypothesis that HBV-induced liver disease is mediated both as a direct effect of cytotoxicity associated with viral protein synthesis, especially aberrant overexpression of the large envelope protein, and as a consequence of immune reactivity to viral envelope antigens expressed on infected hepatocytes. Obviously, the immunogenicity of the remaining HBV-encoded antigens with respect to the induction of CTL and the in vivo susceptibility of the corresponding transgenic hepatocytes to destruction by such CTL against a variety of MHC backgrounds should be assessed to understand fully the potential contribution of the response to each HBV antigen with respect to viral clearance and liver cell injury in viral hepatitis.

5 HBV Gene Expression During the Acute Phase Response

Much of what is known about expression of the HBV genome is derived from studies in which the viral genes and regulatory elements have been introduced into mammalian cells which may bear little resemblance to the primary hepatocyte in which the virus normally exists. Furthermore, although viral hepatitis is an inflammatory process, nothing is known about the influence of inflammatory cells and their products on HBV expression by the liver cell. This is important because inflammation characteristically induces profound alterations in the expression of several hepatocellular proteins (acute phase response) including C-reactive protein, fibrinogen, serum anyloid A, alpha-2-macroglobulin, haptoglobin, albumin, metallothionein, alpha-1-antitrypsin, C3, etc. The acute phase response is mediated by inflammatory cell cytokines, most notably tumor necrosis factor, interleukin-1, interleukin-6, and gamma-interferon (FEY and GAULDIE 1989), which regulate hepatocellular gene expression at the transcriptional and posttranscriptional levels (FEY and GAULDIE 1989).

Since the HBV genome is exposed to the same inflammatory mediators as endogenous hepatocellular genes during acute and chronic hepatitis, it is possible that HBV gene expression might also be modulated during the acute phase response, and that such changes could influence the evolution of the disease. Clearly, this is an important aspect of the virus-host interaction that deserves attention.

Since the acute phase response can be induced experimentally by the administration of bacterial lipopolysaccharide (LPS), we (P. GILLES and F.V. CHISARI, unpublished observations) analyzed HBV gene expression in transgenic mice following the intraperitoneal administration of LPS under conditions known to induce an acute phase response (GLIBETIC and BAUMANN 1986). HBV-specific gene expression (steady state mRNA) was substantially decreased after LPS administration concomitant with an increase in transcripts corresponding to the acute phase proteins SAA and alpha-2-macroglobulin. The molecular basis for these effects is currently under investigation.

Because of the implications of these observations with respect to hepatocellular viral gene expression in the inflamed liver and the derivative implications with respect to the pathogenesis of acute hepatitis, chronic hepatitis, and the healthy carrier state, this area warrants further investigation.

6 Conclusions

Despite recent advances, we are left with major areas of ignorance with respect to the biology and pathogenesis of HBV that the transgenic mouse model can be used to resolve. Its great strength rests in the ability to express each of the viral genes individually and to explore the tissue specificity of every viral regulatory element in all of the organs of the living animal. Indeed, the individual functions of the nucleocapsid, polymerase, and X ORFs remain to be examined in this model. The consequences of an immune response to each of the viral gene products can be studied in this system such that the relative importance of each of the viral antigens and the individual components of the immune response can be assessed in terms of viral clearance, liver cell injury, and hepatocellular carcinoma. The mechanisms responsible for immunological tolerance can be probed in an effort to understand the basis for the establishment of the chronic carrier state in humans. An important objective is the introduction of the HBV receptor into transgenic mice under the control of liver-specific control sequences to permit the creation of the ultimate animal model of HBV infection and the examination of issues relating to vaccine development, antiviral therapy, viral immunobiology, and pathogenesis that are otherwise unapproachable at this time.

Acknowledgments. I thank Takashi Moriyama Patrick Gilles, John Shutter, Pierre Filippi, Carl Pinkert, Richard Palmiter, and Ralph Brinster for collaboration in the production and analysis of transgenic mice. I thank Darlene Corrigan for manuscript preparation. This research was supported by NIH grants CA40489, CA38635, AI20001, AI20720, HD17321, HD0912, and HD07155 and by a North Atlantic Treaty Organization Grant for International Collaboration (675/84). This is publication number 6246-MEM from the Department of Molecular and Experimental Medicine, Research Institute of Scripps Clinic, La Jolla, CA, USA.

References

Alter HJ (1988) Transfusion-associated non-A, non-B hepatitis. The first decade. In: Zuckerman AJ (ed) Viral hepatitis and liver disease. Liss, New York, pp 537–542

Araki K, Miyazaki J, Tsurimoto T, Inomoto T, Iwanaga T, Matsubara K, Yamamura K (1989a) Demethylation by 5-azacytidine results in the expression of hepatitis B virus surface antigen in transgenic mice. Jpn J Cancer 80: 295–298

Araki K, Miyazaki J, Hino O, Tomita N, Chisaka O, Matsubara K, Yamamura K (1989b) Expression and replication of hepatitis B virus genome in transgenic mice. Proc Natl Acad Sci USA 86: 207–211

Babinet C, Farza H, Morello D, Hadchouel M, Pourcel C (1985) Specific expression of hepatitis B surface antigen (HBsAg) in transgenic mice. Science 230: 1160–1163

Bulla G, Siddiqui A (1988) Negative regulation of the hepatitis B virus pre-S1 promoter by internal DNA sequences. In: Abstracts of papers presented at the 1988 meeting on molecular biology of hepatitis B Virus, La Jolla. University of California, San Diego Aug 7–11, 1988; p 64

Burk RD, DeLoia JA, ElAwady MK, Gearhart JD (1988) Tissue preferential expression of the hepatitis B virus (HBV) surface antigen gene in two lines of HBV transgenic mice. J Virol 62: 649–654

Buscher M, Reiser W, Will H, Schaller H (1985) Transcripts and the putative RNA pregenome of duck hepatitis B virus implications for reverse transcription. Cell 40: 717–724

Carlson J, Eriksson S (1985) Chronic "cryptogenic" liver disease and malignant hepatoma in intermediate alpha-1-antitrypsin deficiency identified by a PiZ-specific monoclonal antibody. Scand J Gastroenterol 20: 835–842

Chisari FV, Pinkert CA, Milich DR, Filippi P, McLachlan A, Palmiter RD, Brinster RL (1985) A transgenic mouse model of the chronic hepatitis B surface antigen carrier state. Science 230: 1157–1160

Chisari FV, Filippi P, McLachlan A, Milich DR, Riggs M, Lee S, Palmiter RD, Pinkert CA, Brinster RL (1986) Expression of hepatitis B virus large envelope polypeptide inhibits hepatitis B surface antigen secretion in transgenic mice. J Virol 60: 880–887

Chisari FV, Filippi P, Buras J, McLachlan A, Popper H, Pinkert CA, Palmiter RD, Brinster RL (1987) Structural and pathological effects of synthesis of hepatitis B virus large envelope polypeptide in transgenic mice. Proc Natl Acad Sci USA 84: 6909–6913

Chisari FV, Klopchin K, Moriyama T, Pasquinelli C, Dunsford HA, Sell S, Pinkert CA, Brinster RL, Palmiter RD (1989a) Molecular pathogenesis of hepatocellular carcinoma in hepatitis B virus transgenic mice. Cell 59: 1145–1156

Chisari FV, Ferrari C, Mondelli MU (1989b) Hepatitis B virus structure and biology. Microbiol Pathog 6: 311–325

Colgrove R, Simon G, Ganem D (1989) Transcriptional activation of homologous and heterologous genes by the hepatitis B virus X gene product in cells permissive for viral replication. J Virol 63: 4019–4026

DeLoia JA, Burk RD, Gearhart JD (1989) Developmental regulation of hepatitis B surface antigen expression in two lines of hepatitis B virus transgenic mice. J Virol 63: 4069–4073

Dunsford HA, Sell S, Chisari FV (1990) Hepatocarcinogenesis due to chronic liver cell injury in hepatitis B virus transgenic mice. Cancer Research 50: 3400–3407

Dycaico MJ, Grant SGN, Felts K, Nichols K, WS, Geller SA, Hager JH, Pollard AJ, Kohler SW, Short HP, Jirik FR, Hanahan D, Sorge JA (1988) Neonatal hepatitis induced by alpha-1-antitrypsin. A transgenic mouse model. Science 242: 1409–1412

Eble BE, Lingappa VR, Ganem D (1986) Hepatitis B surface antigen an unusual secreted protein initially synthesized as a transmembrane polypeptide. Mol Cell Biol 6: 1454–1463

Eble BE, MacRae DR, Lingappa VR, Ganem D (1987) Multiple topogenic sequences determine the transmembrane orientation of hepatitis B surface antigen. Mol Cell Biol 7: 3591–3601

El-Ghor MAA, Burk RD (1989) DNase I hypersensitive site maps to the HBV enhancer. Virology 172: 1–10

Enders GH, Ganem D, Varmus H (1985) Mapping the major transcripts of ground squirrel hepatitis virus the presumptive template for reverse transcriptase is terminally redundant. Cell 42: 297–308

Enders GH, Ganem D, Varmus HE (1987) 5′-terminal sequences influence the segregation of ground squirrel hepatitis virus RNAs into polyribosomes and viral core particles. J Virol 61: 35–41

Eriksson S, Carlson J, Velez R (1986) Risk of cirrhosis and primary liver cancer in alpha-1-antitrypsin deficiency. N Engl J Med 314: 736–739

Farza H, Salmon AM, Hadchouel M, Moreau JL, Babinet C, Tiollais P, Pourcel C (1987) Hepatitis B surface antigen gene expression is regulated by sex steroids and glucocorticoids in transgenic mice. Proc Natl Acad Sci USA 84: 1187–1191

Farza H, Hadchouel M, Scotto J, Tiollais P, Babinet C, Pourcel C (1988) Replication and gene expression of hepatitis B virus in a transgenic mouse that contains the complete viral genome. J Virol 62: 4144–4152

Fey GH, Gauldie J (1989) The Acute phase response of the liver in inflammation. In: Popper H, Schffman F (eds) Progress in liver diseases, Vol. 9 Saunders, Philadelphia

Gerber MA, Hadziyannis S, Vissoulis C, Schaffner F, Paronetto F, Popper H (1974) Electron Microscopy and Immunoelectronmicroscopy of cytoplasmic hepatitis B antigen in hepatocytes. Am J Pathol 75: 489–502

Glibetic M, Bauman H (1986) Influence of chronic inflammation on the level of mRNA for acute-phase, reactants in the mouse liver. J Immunol 137: 1616–1622

Hadchouel M, Farza H, Simon D, Tiollais P, Pourcel C (1987) Maternal inhibition of hepatitis B surface antigen gene expression in transgenic mice correlates with de novo methylation. Nature 329: 454–456

Hino O, Nomura K, Ohtake K, Kawaguchi T, Sugano H, Kitagawa T (1989) Instability of integrated hepatitis B virus DNA with inverted repeat structure in a transgenic mouse. Cancer Genet Cytogenet 37: 273–278

Kaneko S, Miller RH (1988) X-region-specific transcript in mammalian hepatitis B virus-infected liver. J Virol 62: 3979–3984

Korba BE, Wells F, Tennant BC, Yoakum GH, Purcell RH, Gerin JL (1986) Hepadnavirus infection of peripheral blood lymphocytes in vivo woodchuck and chimpanzee models of viral hepatitis. J Virol 58: 1–8

Korba BE, Cote PJ, Wells FV, Baldwin B, Popper H, Purcell RH, Tennant BC, Gerin JL (1989) Natural History of woodchuck hepatitis virus infections during the course of experimental viral infection. Molecular virologic features of the liver and lymphoid tissues. J Virol 63: 1360–1370

Lieber CS, Garro A, Leo MA et al. (1986) Alcohol and cancer, Hepatology 6: 1005–1009

Limmer J, Fleig WE, Leupold D, Bittner R, Ditschuneit H, Beger H-G (1988) Hepatocellular carcinoma in type I glycogen storage disease. Hepatology 8: 531–537

Marquardt O, Heermann KH, Seifer M, Gerlich WH (1987) Cell type specific expression of pre S 1 antigen and secretion of hepatitis B virus surface antigen. Brief report. Arch Virol 96: 249–256
McLachlan A, Milich DR, Raney AK, Riggs MG, Hughes JL, Sorge J, Chisari FV (1987) Expression of hepatitis B virus surface and core antigens Influences of pre-S and precore sequences. J Virol 61: 683–780
Melia WM, Johnson PJ, Neuberger J (1984) Hepatocellular carcinoma in primary biliary cirrhosis detection by alpha-fetoprotein estimation. Gastroenterology 87: 660–663
Moller B, Uwe H, Stemerowicz R, Henze G, Gelderblom H (1989) HBcAg expressed on the surface of circulating dane particles in patients with hepatitis B virus infection without evidence of anti-HBc formation. Hepatology 10: 179–185
Molnar-Kimber KL, Jarocki-Witek V, Dheer SK, Vernon SK, Conley AJ, Davis AR, Hung PP (1988) Distinctive properties of the hepatitis B virus envelope proteins. J Virol 62: 407–416
Moriyama T, Guilhot S, Klopchin K, Moss B, Pinkert CA, Palmiter RD, Brinster RL, Kanagawa O, Chisari FV (1990) Immunobiology and pathogenesis of hepatocellular injury in hepatitis B virus transgenic mice. Science 248: 361–364
Niederau C, Fischer R, Sonnenberg A et al. (1985) Survival and causes of death in cirrhotic and in noncirrhotic patients with primary hemochromatosis. N Engl J Med 313: 1256–1262
Ou J, Rutter WJ (1985) Hybrid hepatitis B virus-host transcripts in a human hepatoma cell. Proc Natl Acad Sci USA 82: 83–87
Ou JH, Rutter WJ (1987) Regulation of secretion of the hepatitis B virus major surface antigen by the Pre-S1 protein. J Virol 61: 782–786
Ou JH, Yeh CT, Yen TSB (1989) Transport of hepatitis B virus precore protein into the nucleus after cleavage of its signal peptide. J Virol 63: 5238–5243
Patzer EJ, Nakamura GR, Yaffe A (1984) Intracellular transport and secretion of hepatitis B surface antigen in mammalian cells. J Virol 51: 346–353
Patzer EJ, Nakamura GR, Simonsen CC, Levinson AD (1986) Intracellular assembly and packaging of hepatitis B surface antigen particles occur in the endoplasmic reticulum. J Virol 58: 884–892
Pourcel C (1989) Souris transgeniques pour le genome du virus de l'hepatite B. Medecine/Sciences 5: 626–636
Rall LB, Standring DN, Laub O, Rutter WJ (1983) Transcription of hepatitis B virus by RNA polymerase II. Mol Cell Biol 3: 1766–1773
Schlicht HJ, Schaller H (1989) The secretory core protein of human hepatitis B virus is expressed on the cell surface. J Virol 63: 5399–5404
Seto E, Yen TS, Peterlin BM, Ou JH (1988) Trans-activation of the human immunodeficiency virus long terminal repeat by the hepatitis B virus X protein. Proc Natl Acad Sci USA 85: 8286–8290
Siddiqui A, Jameel S, Mapoles J (1986) Transcriptional control elements of hepatitis B surface antigen gene. Proc Natl Acad Sci USA 83: 566–570
Simon K, Lingappa VR, Ganem D (1988) Secreted hepatitis B surface antigen polypeptides are derived from a transmembrane precursor. J Cell Biol 107: 2163–2168
Spandau DF, Lee CH (1988) Transactivation of viral enhancers by the hepatitis B virus X protein. J Virol 62: 427–434
Standring DN, Ou JH, Rutter WJ (1986) Assembly of viral particles in Xenopus oocytes presurface-antigens regulate secretion of the hepatitis B viral surface envelope particle. Proc Natl Acad Sci USA 83: 9338–9342
Tur-Kaspa R, Burk RD, Shaul Y, Shafritz DA (1986) Hepatitis B virus contains a glucocorticoid-responsive element. Proc Natl Acad Sci USA 83: 1627–1631
Twu J-S, Schloemer RH (1987) Transcriptional trans-activating function of hepatitis B virus. J Virol 61: 3448–3453
Twu JS, Chu K, Robinson WS (1989) Hepatitis B virus X gene activates kB-like enhancer sequences in the long terminal repeat of human immunodeficiency virus 1. Proc Natl Acad Sci USA 86: 5168–5172
Will H, Reiser W, Weimer T, Pfaff E, Buscher M, Sprengel R, Cattaneo R, Schaller H (1987) Replication strategy of human hepatitis B virus. J Virol 61: 904–911
Wollersheim M, Debelka U, Hofschneider PH (1988) A transactivating function encoded in the hepatitis B virus X gene is conserved in the integrated state. Oncogene 3: 545–552
Yamamura K, Tsurimoto T, Ebihara T, Kamino K, Fujiyama A, Ochiya T, Matsubara K (1987) Methylation of hepatitis B virus DNA and liver-specific suppression of RNA production in transgenic mouse. Jpn J Cancer Res 78: 681–688

Cellular and Molecular Mechanisms of Hepatocarcinogenesis Associated with Hepadnavirus Infection

C. E. ROGLER

Liver Research Center, Department of Medicine, Albert Einstein College of Medicine, 1300 Morris Park Avenue, Bronx, NY 10401, USA

Current Topics in Microbiology and Immunology, Vol. 168

1 Fundamental Concepts in Hepatocarcinogenesis

1.1 Activation of Carcinogens, Initiation, and Promotion in the Liver

A large number of inorganic and organic chemicals act as carcinogens in the liver. Most of these are not active until they are metabolized into short-lived electrophilic forms by the mixed function oxidase systems, e.g., P450 and P448, which is most abundant in the liver. The process of carcinogenesis is initiated when an electrophilic form binds to cellular macromolecules, most importantly DNA, and causes a permanent alteration in the nucleotide sequence. Some of the cells that have been altered in this manner are said to be "initiated", meaning that they have acquired characteristics which may be expressed in cancer cells (SCHERER 1989; FARBER 1973; SELL et al. 1987).

The molecular events following initiation are not well understood. Only a small fraction of the hepatocytes in a liver exposed to a carcinogen become initiated, but these cells represent an important pool of precursor cells which are susceptible to additional changes. Moreover, the number of initiated cells increases with the dose and time of exposure to carcinogen and by proliferation of the initiated cells themselves.

Clonal expansion of initiated cells is enhanced by tumor promoters and inhibited by anticarcinogens. Clonal expansion also increases the probability that a cell will undergo a second genetic change. If this genetic change imparts a selective advantage to the cell, it will again clonally expand to give rise to a population of cells which are candidates for a third genetic change. In this way, properties such as independence of specific growth factors and metastasizing capacity develop in a stepwise manner (SCHERER 1984). Thus, the term promotion refers to the process in which cancer is promoted in cells which have first been initiated (FARBER 1980, 1984; SELL et al. 1987; PITOT and SINCA 1980).

1.2 Cellular Origins of Primary Hepatocellular Carcinoma

Different hepatocarcinogens and carcinogenic regimens produce markedly different cellular reactions in the liver. In the classic models (FARBER 1973; SCHERER 1984; SELL et al. 1987), small collections of altered hepatocytes, characterized by more basophilic staining properties indicative of active growth, arise in the liver in the initial stages of carcinogenesis. This is followed by the appearance of large clusters (nodules) of altered hepatocytes that compress the adjacent normal-appearing liver cells. These nodules increase in size during the carcinogenic process, and their morphological appearance suggests that they are precursors of hepatocellular carcinoma (HCC). The appearance of further altered foci of cells, or "nodules within nodules," has been taken as evidence for carcinogenic progression (FARBER 1984).

A number of different regimens for exposing rats to chemical hepatocarcinogens have been developed which have enabled investigators to obtain a better understanding of which cells are the first to respond to treatment with a carcinogen. Electron microscopic autoradiography has shown that the first cells to proliferate, 1-3 days after carcinogen treatment, are nondescript, periportal, small "oval cells" (SELL and SALMAN 1984). These cells have little cytoplasm and no morphological features which identify them with a particular cell type (SELL et al. 1981). Certain carcinogen regimens promote the accumulation of large numbers of oval cells (SELL 1978), while others stimulate only moderate oval cell accumulation. Pulse chase labeling studies using [^{3}H] thymidine to label replicating DNA have shown that label in oval cell nuclei is preserved in the nuclei of hepatocytes present in regenerative nodules (EVARTS et al. 1987). These studies lend strong support to the notion that oval cells are true liver stem cells which give rise to cells in the hepatocyte lineage leading to regeneration nodules. Oval cells also share common antigens with bile ductular cells and can differentiate along the bile ductular cell lineage (HIXSON and ALLISON 1985; SELL and SALMAN 1984). The factors which control the transition from stem cell to mature hepatocyte or bile duct are unknown. However, monoclonal antibodies which are markers for bile ducts, oval cells, transitional cells, and fully differentiated hepatocytes have been developed and are providing valuable tools to study cell lineages in the liver (HIXON and ALLISON 1985; DUNSFORD and SELL 1989; DUNSFORD et al. 1989; HIXON et al. 1990).

Since dividing stem cells are presumably susceptible targets for carcinogens, it would be predicted that, as a result of mutagenesis of these cells, cholangiocarcinoma as well as HCC would occur in the liver of rats treated with carcinogens and in the liver of hepadnavirus carriers (SELL et al. 1987; SUMMERS et al. 1978). Tumorigenesis along the bile ductular cell lineage is observed initially as ductular hyperplasia, then cholangiofibrosis, and finally cholangiocarcinoma. Cholangiocarcinoma is much rarer than HCC, and this may have to do with the size of the precursor cell population as well as other unknown factors controlling transformation of bile ductular cells.

An illustration of the proposed cellular pathways to primary HCC is presented in Fig. 1. The classic pathway hypothesizes that hepatocytes are altered and proliferate to produce foci which lead to nodules and eventually cancer. However, recent evidence strongly suggests that the predominant cellular pathway to HCC begins with oval cell rather than hepatocyte proliferation, followed by differentiation of these initiated oval cells to form nodules and atypical hyperplastic zones in the liver (FARBER 1984; LOMBARDI 1982; HIXON et al. 1990). These precancerous lesions represent independent populations of cells at various stages in carcinogenic progression, and only a small percentage of them progresses to cancer.

In normal liver, hepatocyte turnover is slow, and primary HCC is very rare. In carcinogenic regimens, a continuous or repeated mitogenic stimulus is required, in conjunction with carcinogen treatment, for progression to HCC to occur with significant frequency. However, the combination of mitogenic stimulus and

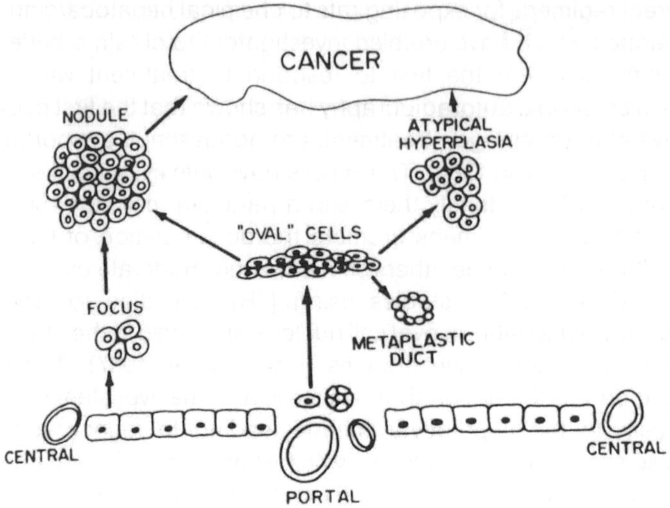

Fig. 1. Possible cellular lineages in experimental hepatocarcinogenesis in rats (adapted from SELL et al. 1987). Possible relationships between putative premalignant cellular changes in the livers of rats exposed to chemical carcinogens and the malignant tumor that eventually appears are depicted. Liver stem cells present in the portal region proliferate and appear as "oval" cells which differentiate to form atypical hyperplasia and/or nodules. These progress to form hepatocellular carcinoma. Carcinoma may also arise from normal hepatocytes that become altered and proliferate to form a small "focus" that may progress to nodules and then to cancer

carcinogen treatment transform the liver into a highly susceptible target for carcinogenesis. In the remainder of this review we will discuss how hepadnaviruses provide both the mitotic stimulus and the carcinogenic agent which are necessary for the development of HCC.

2 Hepadnavirus Infection and HCC

2.1 Acute HBV Infection in Relation to the Carcinogenic Mechanisms

During acute HBV infection, both degenerative and regenerative changes coexist. In acutely infected individuals the development of a complete immune response is sufficient to eliminate viral replication completely, allowing recovery to proceed. During the recovery phase, liver regeneration is prominent and is manifested by anisonucleosis, bi- and trinucleate cells, and mitoses (HOLLINGER 1985a).

In terms of carcinogenic models, the liver injury and regeneration which occur during acute HBV infection can serve to generate initiated hepatocytes, since acutely infected individuals have an elevated risk of developing HCC. The number of initiated hepatocytes produced probably depends on the severity of the hepatitis and the length of time the infection persists. The presence of initiated hepatocytes also represents an irreversible step towards susceptibility to HCC. However, the initiated hepatocytes will not progress toward malignancy without the further action of tumor promoters. Thus, individuals experiencing only acute infections are still at much lower risk than those which develop persistent infection (BEASLEY et al. 1981).

2.2 Persistent HBV Infection and Hepatocarcinogenesis

It was previously thought that acute HBV infection progresses to persistent infection in 5%–10% of adult HBV patients (HOLLINGER 1985a; BEASLEY et al. 1981). The use of Polymerase Chain Reaction (PCR) methodology has shown that HBV infection may persist in a much greater percentage of individuals. The course of chronic liver disease caused by persistent HBV infection is quite varied. In the case of asymptomatic HBV carriers, liver disease may be barely detectable for years. Alternatively, persistent infection may cause a relentless progression to cirrhosis and death (HOLLINGER 1985b). In regard to carcinogenic risk, even these persistently infected individuals, who may be completely asymptomatic, are probably exposed to both a continuous carcinogenic agent (the hepadnavirus) and a mitogenic stimulus (generated by a limited immune response) placing them at very high risk of developing HCC. In fact, male HBV carriers have a 40% lifetime risk of HCC (BEASLEY et al. 1981). The results do not, however, rule out the possibility that some carriers are actually at very low risk because of the very low level of mitogenic stimulus.

In humans, several stages in the progression from mild persistent hepatitis to chronic active hepatitis (CAH) and finally to HCC have been observed. In most HBV carriers, replication gradually ceases over the course of many years while hepatocytes continue to express viral envelope proteins. RAIMONDO and colleagues (RAIMONDO et al. 1988) have also found that viral DNA replication may continue in some patients in the absence of virus production. These findings are all consistent with a breakdown in the ability of hepatocytes to support a critical event in virus maturation or gene expression subsequent to viral DNA synthesis (MASON and TAYLOR 1989) and suggest that infection-associated injury induces dramatic changes in the hepatocyte population, perhaps including a selection for hepatocytes that are no longer able to support virus replication.

In fact, in many individuals active virus replication eventually either ceases or declines to very low levels, with concurrent loss of HBeAg from the serum and the appearance of HBe-specific immunoglobulins (chapter by D. GANEM, this volume). The observation of increased liver disease activity during this transition period suggests a major loss of infected hepatocytes, and their replacement via regeneration with hepatocytes that may be nonpermissive for virus replication. This regenerative activity could, of course, greatly expand the pool of cancer precursor cells, placing the individual at much greater risk for HCC than that which existed immediately prior to seroconversion.

2.3 Animal Models of Hepadnavirus Infection

Four hepadnaviruses have been identified in addition to HBV, including woodchuck hepatitis virus (WHV) (SUMMERS et al. 1978), ground squirrel hepatitis virus (GSHV) (MARION et al. 1980), duck hepatitis B virus (DHBV) (MASON et al. 1980), and heron hepatitis B virus (HHBV) (SPRENGEL et al. 1988). Of these animal viruses, WHV infections are being most actively studied to gain insights into HBV infections in humans. Therefore, WHV infections and their associated pathogenic consequences will be discussed to the greatest extent in the following sections.

2.3.1 Woodchuck Hepatitis Virus

Acute WHV infection in the liver of woodchucks is characterized by focal eosinophilic changes of the cytoplasm of hepatocytes. Activation of sinusoidal cells occurs, and portal tracts show extensive infiltration by lymphocytes. Lymphocytes and plasma cells also infiltrate the liver parenchyma but to a lesser extent than in the portal tracts (POPPER et al. 1987). At the height of acute lesions, proliferated bile ductules are present in the periportal areas (POPPER et al. 1987; KORBA et al. 1987). Within the context of carcinogenic models, these acute lesions may provide the mitogenic stimulus needed to propagate mutation in initiated cells that may be present in the liver as a result, perhaps, of events unrelated to virus infections (but see below). The existence of proliferating bile ductules and newly regenerated hepatocytes in acutely infected woodchuck liver (POPPER et al. 1987) is consistent with the possibility that liver stem cell proliferation is responsible for part, if not all, of the restoration of liver mass in animals recovering from acute infection.

Several cases of HCC have indeed been reported in woodchucks which experienced acute WHV infection followed by seroconversion to WHc- and WHs-specific antibodies (KORBA et al. 1987). The specific mechanism by which WHV served as a carcinogenic agent in these cases is unknown. However, the tumors in these animals contained WHV DNA integrations, and since integration is a rare event during acute infection (see later section) these results suggest that cells containing integrated WHV may have been selectively targeted, or placed at higher risk, for future malignant transformation possibly involving N-*myc* activation (see Sect. 5.2).

Persistent WHV infection is commonly characterized by a state of mild lymphocytic infiltration in the portal tracts known as portal hepatitis (PH) (POPPER et al. 1987; KORBA et al. 1987). In PH, hepatocytes which are analogous to "ground glass" cells in HBV carriers are frequent (see chapter by F.V. CHISARI, this volume, for review). This minimal state in woodchucks corresponds to the human "healthy" HBsAg carrier state. Mild PH develops most frequently in neonatal rather than adult infections of woodchucks and presents a high risk of HCC (POPPER et al. 1987; SNYDER and SUMMERS 1980); adult woodchucks usually develop acute infections which are resolved. In the most carefully controlled study conducted so far, 8 of 8 (100%) captive woodchucks infected at birth with WHV and maintained in captivity developed HCC after 17–36 months of

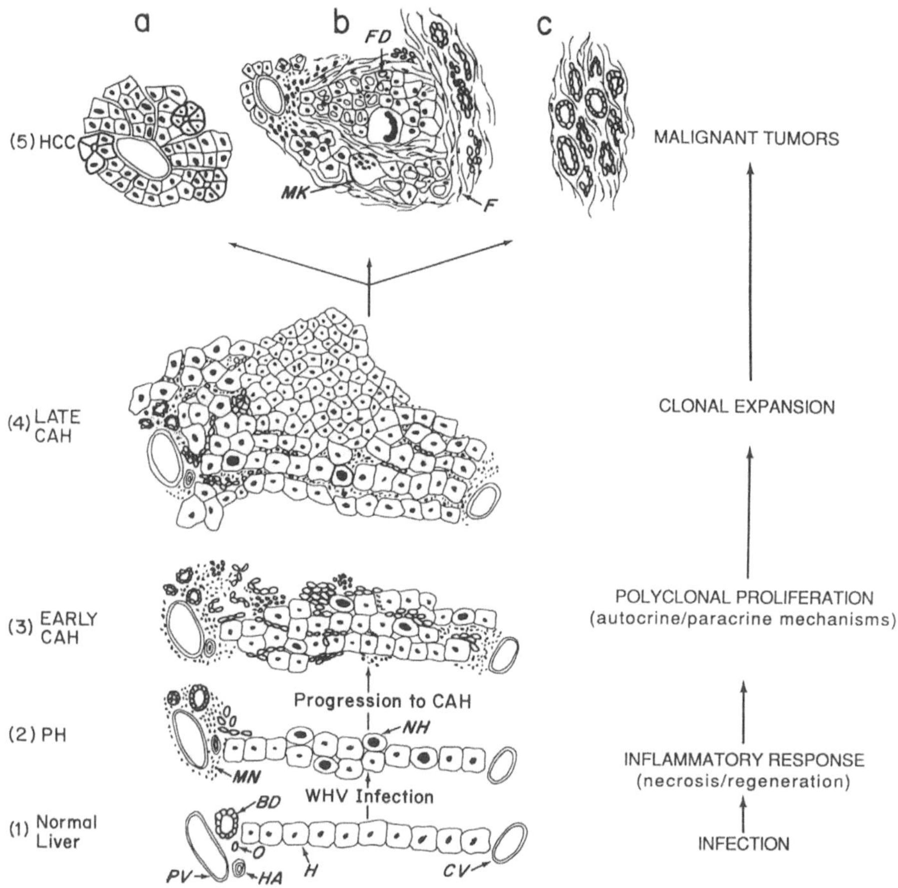

Fig. 2 a–c. The proposed progressive stages of WHV infection leading to the generation of biochemically and histologically heterogeneous primary hepatocellular carcinoma (HCC) in woodchucks. *O*, oval cell; *H*, hepatocyte; *BD*, bile duct; *CV*, central vein; *PV*, portal vein; *HA*, hepatic artery; *MN*, mononuclear cell; *F*, fibroblast; *MK*, megakaryocyte; *FD*, fatty degeneration; *NH*, necrotic hepatocyte. *1*, Schematic illustration of normal liver architecture comprising portal tracks (*left*) connected by single cell plates of hepatocytes and central hepatic veins (*right*). *2*, After establishment of WHV persistent infection, portal hepatitis (*PH*) develops. *3*, Portal hepatitis progresses to chronic active hepatitis (*CAH*). Oval cell proliferation is prominent. *4*, Precancerous nodules develop throughout the liver parenchyma as oval cell proliferation and an inflammatory response continue. *5*, Sustained by the dynamics of protracted infection, inflammation, cell necrosis, and growth factor production, heterogeneous types of tumors arise. Illustrated are; (**a**) a highly differentiated "adult type" HCC, (**b**) a poorly differentiated, anaplastic, "fetal type" HCC with extensive extramedullary hematopoesis, and (**c**) a rare cholangiocarcinoma

persistent infection. This represents the highest reported incidence of HCC obtained following insult with any carcinogenic agent (POPPER et al. 1987).

During mild PH, oval cells are present in the portal tracts of WHV-infected woodchucks in low but elevated numbers compared with their very rare occurrence in normal liver (FU et al. 1988). However, a period of necroinflammation is almost always observed in the precancerous stages preceding and concurrent with the development of HCC (POPPER et al. 1987). During this precancerous period oval cell proliferation may become extensive, with focal accumulations of oval cells appearing in the centrilobular regions of the liver along with cords of cells exhibiting transitional phenotypes (FU et al. 1988).

A model to explain the possible relationship of reparative cellular proliferation and the eventual occurrence of HCCs with specific biochemical and histological phenotypes is presented in Fig. 2. The model proposes that once WHV persistent infection is established, PH ensues, and several pathological reactions occur on a self-perpetuating basis. These include necrosis and regeneration of hepatocytes, an inflammatory reaction primarily limited to the portal tracts, and mild proliferation of bile ducts (Fig. 2, line 2). As the disease progresses to CAH, involvement of the rest of the hepatic lobule becomes evident (lobular hepatitis) (Fig. 2, line 3). Oval cells, either isolated or in clusters, can be found along the sinusoids from portal tracts to the midzone and centrilobular areas of the hepatic lobule. From within this milieu of cells, precancerous nodules develop (Fig. 2, line 4), and selected nodules progress to more malignant phenotypes through the autocrine and paracrine activities of growth factors (see Sect. 5). In the final stage, tumors with various phenotypes, ranging from highly differentiated to poorly differentiated HCCs arise in the liver (Fig. 2, line 5).

Therefore, one explanation for the heterogeneity of tumor phenotypes which arise in the liver is that transitional cells which have not become fully differentiated hepatocytes are present in precancerous liver. The further differentiation of these cells is likely to be arrested as a result of, or as part of, malignant transformation since arrested differentiation is a common feature of many types of tumor cells (HEARD et al. 1984). Thus, the final phenotype of the tumor most likely reflects the state of differentiation of the precursor cell at the time of transformation. According to this model, transformation of transitional hepatocytes generates less differentiated, more fetal-like tumors, whereas transformation of cells further along the hepatocyte lineage produces a more highly differentiated tumor.

2.3.2 Ground Squirrel Hepatitis Virus and Duck Hepatitis B Virus

Experimental infection of ground squirrels with GSHV results in either (1) self-limited or transient acute infection, (2) a persistent infection, or (3) antibody production without detectable viral replication (MARION et al. 1983). The predominant histologic reaction during persistent infection is a mild hepatitis. Some portal tracts are expanded by infiltration with mononuclear inflammatory cells which surround the bile ducts and ductules. Proliferated bile ductules are present, and inflammation sometimes extends into the surrounding parenchyma, replacing hepatocytes in a few places (MARION et al. 1983).

HCC takes longer to develop in ground squirrels than woodchucks, perhaps reflecting the fact that a mild hepatitis is retained for longer periods, rather than progressing as in the woodchuck (MARION et al. 1986). Nonetheless, in older ground squirrels, with persistent GSHV infection, there is a very high rate of progression to HCC, similar to that in woodchucks (MARION et al. 1986). HCC may also develop in ducks as a result of persistent infection with DHBV (YOKOSUKA et al. 1985). Histologic analysis of persistent infection in ducks reveals an even milder hepatitis than in ground squirrels (OMATA et al. 1983). Consistent with this, HCC may take even longer to develop in persistently infected ducks than in ground squirrels (YOKOSUKA et al. 1985).

3 Mechanisms of Hepatocellular Injury —Maintenance of the Mitogenic Stimulus

Since cell death and regeneration are essential parts of the carcinogenic process, it is important to understand the mechanisms by which viral infection promotes these activities. Since most infected hepatocytes appear healthy during persistent infection, and viral replication can be maintained in tissue culture without cell death, it is evident that hepadnaviruses are not normally cytotoxic to their host cells. Hepatocellular injury is therefore mediated by the host immune system and under certain circumstances by abnormal production of various viral proteins.

3.1 Immunopathogenesis of Persistent HBV Infection

The primary issues regarding immunopathogenesis during persistent HBV infection involve the role of the cytotoxic T-lymphocyte (CTL) response and of the B-cell-mediated immune response as mediators of liver damage. Studies using mouse models have provided significant insights into the role of T-cell-independent and T-cell-dependent immune responses in the mouse (MILICH and CHISARI 1982; MILICH et al. 1985, 1986, 1987), and these studies provide a basis for further investigations of the role of these responses in woodchucks and humans. The degree to which various viral proteins contribute to the CTL response or to antibody-dependent cellular cytotoxicity is still a matter of active investigation. The CTL response could be directed to several HBV-encoded antigens displayed on the hepatocyte membrane including the envelope, nucleocapsid, polymerase, or even the X polypeptides. The failure to generate antibodies to HBs antigens is believed to be one of the factors required for the maintenance of persistent infection (CHISARI et al. 1989a). Since helper T cells are thought to be involved in both the generation of the CTL response and the B-cell immune response, functional failure at the helper T cell could also be an important factor regulating the degree of pathogenesis during persistent infection (CHISARI et al. 1989a).

3.2 Viral Antigens and Hepatotoxicity

Accumulation of viral antigens in hepatocytes during acute and persistent infections exhibits a wide range of variability. In most cells, viral envelope proteins are rapidly secreted from infected hepatocytes in the form of 22-nm spheres and filaments and Dane particles. In other instances, particularly during chronic infection, many hepatocytes accumulate envelope proteins and display histological and ultrastructural features described as "ground glass" hepatocytes (see chapter by F.V. CHISARI, this volume).

In transgenic mouse lines in which the overexpression of the large envelope protein of HBV, as compared with middle and small proteins, has been specifically induced, an aberrant accumulation of HBsAg particles in the endoplasmic reticulum occurs which eventually kills the hepatocytes, leading to a nonimmunologic, inflammatory response (CHISARI et al. 1987). The liver cell death induced by this HBsAg accumulation is associated at first with the formation of ground glass cells, presumably similar to those in human infections and reflecting a stage leading to cell death. This cell death initiates a complex sequence of events culminating in the development of HCC in all transgenic mice of the lineage (CHISARI et al. 1989b). The continuous regeneration response in the transgenic mouse liver also results in the proliferation of oval cells and microscopic nodular regenerative hyperplasia during the precancerous stages. As in woodchucks, two general histologic subsets of HCC were observed in the mice. These included trabecular, well-differentiated, and mitotically quiet tumors and solid, sheet-like, poorly differentiated, and mitotically active tumors (CHISARI et al. 1989b). These results are consistent with the model proposed earlier that transformation of cells at various stages of differentiation arrests their further differentiation and results in tumors with heterogeneous phenotypes (chapter by F.V. CHISARI, this volume).

In general, cell death and regeneration are believed to be insufficient to produce HCC without the concomitant activity of a carcinogenic agent. Chromosome analysis of cells in precancerous livers and HCCs of the transgenic mouse livers revealed populations of cells with unique aneuploid chromosome contents (CHISARI et al. 1989b). These results suggest that major chromosomal alterations with the potential to disrupt growth regulatory mechanisms occur frequently in precancerous and HCC tissues. It is interesting that, in the transgenic mouse, cell death and the resulting regeneration may be the immediate determinants of chromosomal damage. However, during hepadnavirus infections, viral DNA integration also plays a direct role in generating chromosome abnormalities. In the next section, viral DNA integration and other virally associated controls of gene expression will be discussed as they relate to the carcinogenic process.

4 Hepadnavirus DNA Integration and Hepatocarcinogenesis

4.1 Occurrence of Viral Integration in Precancerous Liver

While hepadnaviruses share many common characteristics with retroviruses, several important features have established them as unique (GUST et al. 1986; MILLER and ROBINSON 1986). One of these is that they do not depend on provirus formation for the production of pregenomic RNA; rather, they utilize covalently closed circular (CCC) viral molecules as the transcriptional template for pregenome RNA and mRNA synthesis. The hepadnaviral reverse transcriptase also contain a RNaseH activity, but it does not have a domain which is homologous to the retroviral integrase activity (RADZIWILL et al. 1980). This is understandable in light of the fact that provirus formation is not part of the hepadnavirus replication cycle (chapter by C. SEEGER et al., this volume).

In spite of the fact that integration is not a part of the viral replication cycle, integrated hepadnaviral DNA was discovered as clonally propagated integrations in HCC from hepadnavirus carriers (SUMMERS et al. 1980; CHAKRABORTY et al. 1980; SHAFRITZ et al. 1981; BRECHOT et al. 1980, 1981; KOSHY et al. 1981; EDMAN et al. 1980). This discovery led to speculation that integrations might participate in hepatocellular transformation through the activation of cellular protooncogenes. This proposal does appear to be one mechanism by which integrations may participate in oncogenesis (Sect. 4.3). Integration has also been linked to more general carcinogenic mechanisms involving chromosome rearrangements (Sect. 4.4). Recently, a previously unexpected consequence of integration, generation of a transactivator of cellular promoters, has been demonstrated (Sect. 4.5). These mechanisms will be discussed in more detail after a brief discussion of the occurrence of integration and integration mechanisms.

In order to determine whether hepadnavirus integrations participate in the initiation phase of carcinogenesis, it is necessary to study integrations in precancerous tissues. Frequent integration does not occur during acute infection or in the early stages of persistent infection as judged by the lack of detectable integrated molecules in acutely infected liver. In one study, a duck genomic library from an acutely infected duck liver, comprising 10 genomic equivalents of DNA, was screened for integrated DHBV DNA. While other genomic clones were isolated from this library, no integrated DHBV genomes were detected (unpublished data, C.E. ROGLER and J. SUMMERS). Thus, it appears that during acute infection less than 1 in 10 hepatocytes acquire an integrated genome.

Studies of the occurrence of integrations in the precancerous stages of chronic infection are very limited because almost all the studies of normal tissues were conducted with individuals in whom HCC had already developed in another part of the liver, especially in studies dealing with HBV integration (SHAFRITZ et al. 1981; KOSHY et al. 1981; CHEN et al. 1982). The clearest demonstration of integration during persistent infection, in the absence of HCC, was the cloning of a WHV integration from a biopsy specimen obtained from a persistently infected woodchuck liver (ROGLER and SUMMERS 1984). After screening of a lambda recombinant DNA library comprising four genome equivalents of woodchuck DNA, one integrated clone was isolated. By this criteria, only one in four hepatocytes in this liver contained an integration, although this is a very rough estimate. HCC did not develop in the woodchuck from which the integration was cloned for over a year after the analysis, at which time the animal died of natural causes.

Most cases of integration in precancerous tissues have been detected at later stages of infection, in which CAH has developed and HCC is also present (SHAFRITZ et al. 1981; KOSHY et al. 1981; YAGINUMA et al. 1987; HADA et al. 1986). In many of these cases two patterns of DNA integrations are observed by Southern blot analysis, specifically (1) integration at random positions throughout the genome producing a smear of hybridization in high molecular weight DNA and (2) specific banding patterns superimposed on a background smear, indicating that populations of clonally expanded cells have arisen within the liver. When individual precancerous nodules were studied, some contained clonal integrations, and others did not. The high frequency of integrations in HCC of humans (SHAFRITZ et al. 1981), woodchucks (SUMMERS et al. 1980; FU et al. 1988), ground squirrels (MARION et al. 1986), and ducks (YOKOSUKA et al. 1985) shows that all these hepadnaviruses can integrate in the genome and supports the notion that integrations play a role in the development of HCC. Some specific mechanisms which have been suggested by cloning experiments will be discussed later.

4.2 A Role for Topoisomerase I in Hepadnavirus Integration

The first major advance in our understanding of the viral integration mechanism was the report of DEJEAN et al. (DEJEAN et al. 1984) which identified the DR1 and DR2 sequences in the viral genome and identified integrations of HBV DNA which occurred at both these sites. The DR1 and DR2 sequences (two 11-bp direct repeats) are the sites of initiation of replication of the minus (DR1) and plus (DR2) strands of the viral DNA and are located at the termini of the minus and plus strands, respectively (Fig. 3; see chapter by C. SEEGER et al., this volume). The terminal redundancy in the minus strand causes the open circular genome to have a forked structure at the 3′ end of the minus strand. The presence of an unpaired DNA strand near DR1 led to the proposal that HBV DNA may integrate by a strand invasion mechanism (SHIH et al. 1987). Alternatively, a single nick in

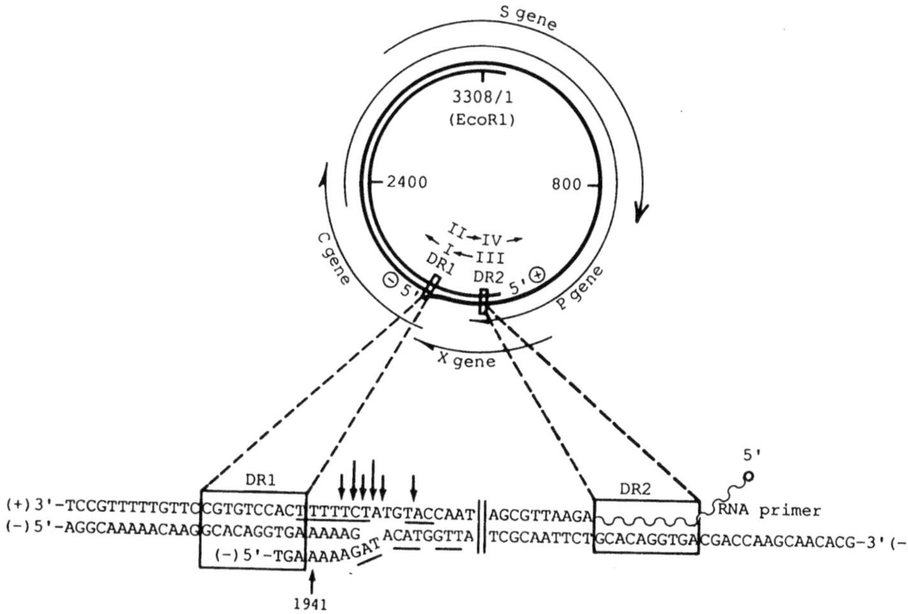

Fig. 3. Illustration of the circular structure of woodchuck hepatitis virus (WHV) Dane particle DNA (referred to as WHV DNA in the text) including the nucleotide sequences in the vicinity of the 5′ ends of the minus (−) and plus (+) strands. The positions of the envelope (*S gene*), core (*C gene*), terminal protein-reverse transcriptase (*P gene*), and (*X gene*) are noted by *arrows* around the circular map. The Roman numerals *I*, *II*, *III*, and *IV* with associated *arrows* refer to prototype integration types I–IV described by Shih et al. (1987). Directly repeated sequences, *DR1* and *DR2*, are denoted within *boxes*. Topoisomerase I cleavage sites mapped by primer extension are illustrated by *arrows* above the WHV DNA strand. Topo I consensus sequences in the region adjacent to the DR1 sequence are denoted by *solid lines* below the WHV (+) and (−) strands. The *vertical solid double line* denotes a break in that portion of the cohesive overlap sequence that was not included in this illustration

the plus strand, opposite the terminal redundancy in the minus strand (near the DR1) or in the minus strand adjacent to the DR2 sequence, might be sufficient to linearize the viral DNA and create molecules which are precursors for integration. Linear molecules would also be formed if the primer of plus-strand DNA synthesis failed to translocate to DR2 (see chapter by C. Seeger et al., this volume). The presence of linear molecules, which might contain activated ends, has been proposed (Nagaya et al. 1987; Tokino et al. 1987), and such linear molecules appear to occur in viral DNA populations.

Since hepadnaviruses do not contain an integrase gene, cellular enzymes must mediate the viral integration pathway. Topoisomerase I (Topo I) is a highly abundant nuclear enzyme which functions in transcription and DNA replication by providing swivels in genomic DNA and allowing movement of transcriptional and replication complexes along the DNA (Champoux 1988; Champoux and Bullock 1988; Gilmour and Elgin 1987; Brill et al. 1987). As part of this mechanism, Topo I generates transient nicks in genomic DNA. It remains

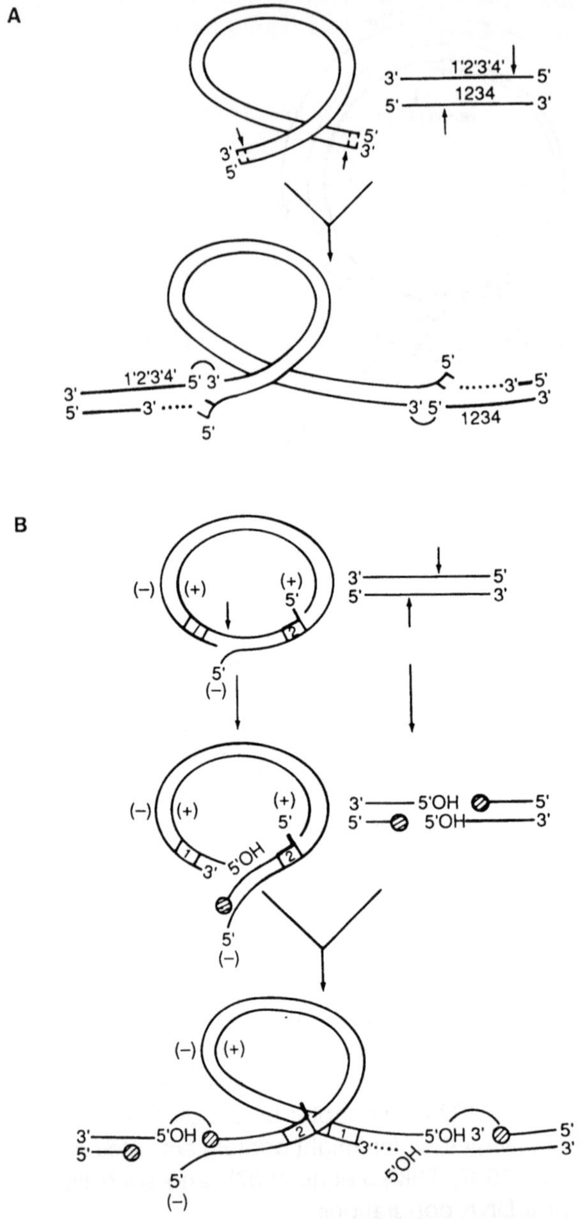

Fig. 4 A, B. Illustration comparing retrovirus integration with the proposed topoisomerase I-mediated hepadnavirus integration mechanism. **A** Retroviral integration. A linear, double-stranded, viral DNA molecule is nicked 2 bp from its 3′ end producing two unpaired bases at each 5′ terminus (double-stranded loop). This molecule is integrated at a cellular DNA site, which is nicked at staggered positions through the action of the viral integrase followed by joining of the 5′ ends of cellular DNA to the 3′ ends of viral DNA. The 2-base 5′ overhang is lost from each end of the viral DNA during integration. The complete integration reaction therefore generates a short direct duplication of cellular DNA at each viral-cell junction. **B** Hepadnavirus integration. Topo I linearizes open circular

covalently bound to DNA via a 3′ phosphotyrosyl bond while the DNA swivels (CHAMPOUX 1977, 1978). If a proper 5′OH acceptor end is provided, these 3′ bound Topo I molecules are capable of linking heterologous DNA fragments via a nonhomologous recombination mechanism (HALLIGAN et al. 1982).

A survey of HBV and WHV sequences across integration sites has shown that preferred Topo I cleavage sites are present at or near 91% of the HBV integration sites cloned from HCC (SCHIRMACHER and ROGLER, unpublished data and Table 1). This value is virtually identical with a previous survey of DNA sequences across sites of illegitimate recombination in which Topo I sites were present in 92% of the cases (KONOPKA 1988). In summary, these data support the notion that Topo I plays a role in HBV integration in vivo.

Recent studies have demonstrated that Topo I cleaves WHV DNA at a cluster of bases near the DR1 sequence in plus-strand DNA and that these sites are located within the cohesive overlap of virion DNA (Fig. 3) (WANG and ROGLER 1991). In addition to cleavage, Topo I can link viral DNA to the ends of heterologous cellular DNA fragments in in vitro reactions (WANG and ROGLER 1991). A new "integration" function to Topo I, in which Topo I mediated integration of WHV virion DNA into an internal site in double-stranded cellular DNA, was also described in this study. The in vitro generated integrations closely resemble in vivo integration in that they show the same end specificity and polarity as in vivo integrations (WANG and ROGLER 1991).

Taking into consideration the Topo I cleavage mechanism, a prototype model for the integration of both ends of a Topo I linearized hepadnaviral molecule has been proposed (WANG and ROGLER et al. 1991), and this model is compared with the integration model of retroviral DNA (GRANDGENETT and MUMM 1990) in Fig. 4. In retroviruses, the integration precursor is a linear DNA molecule. Each strand of the linear molecule is specifically nicked 2 bp from its 3′ end, and the immediate precursor molecule therefore contains two unpaired bases at its 5′ termini. This molecule becomes integrated at a cellular site which is also nicked at positions which produce cohesive ends with 5′ overhangs. The complete integration reaction generates a short direct duplication of cellular DNA due to the separation of the cellular cohesive ends at the integration site (Fig. 4).

The hepadnavirus integration model resembles the retroviral mechanism in some respects but is unique in many others. It is similar in that a linearized viral molecule is the immediate precursor for integration, and cellular DNA is cut at

◄

virion DNA by nicking the plus strand opposite the terminal redundancy in the minus strand (at the fork in virion DNA). It also cleaves cellular DNA at staggered positions. It remains covalently bound to the 3′ phosphate of the viral DNA plus strand and mediates integration, by linking it to a free 5′ OH in the cellular DNA. The opposite end of the plus strand contains a free 5′ OH group which serves as an acceptor site for a 3′ Topo I molecule bound to cellular DNA. The directly repeated (DR1, DR2) sequences in hepadnavirus DNA are noted by *enclosed boxes* with *1* and 2, respectively. *Circles with hash marks* are Topo I molecules. *Arrows* immediately adjacent to DNA molecules are nicking sites. *Thick squiggly line* opposite 2 is the plus-strand RNA primer. *Dots*, regions of the molecules which must be filled in by cellular DNA polymerases staggered positions in opposite strands. However, in the case of hepadnaviruses, integration is usually associated with microdeletions of cellular DNA instead of direct duplications (NAKAMURA et al. 1988; NAGAYA et al. 1987; HINO et al. 1989). One case of a direct duplication has been reported.

According to the hepadnavirus integration model, Topo I is responsible for linearizing circular virion DNA, and it remains bound to the 3′ end of the viral molecule after cleavage and linearization (Fig. 4). These Topo I-containing ends function as "donors" during integration and rearrangement reactions. The opposite end of the viral molecule contains a free 5′OH which can function as an acceptor site for other DNA ends with 3′ bound Topo I molecules. In order for a linearized viral molecule to integrate, it must insert into the reaction site at the moment of Topo I nicking and provide an alternate acceptor or donor end. The complexity of such reactions may account for the low frequency of integration events observed in vivo. On the other hand, reactions which involve intramolecular joining of Topo I-cleaved ends of the same molecule would be expected to occur more often and would generate rearranged molecules. Most integrations cloned from HCC have inversions and deletions, possibly due to Topo I-mediated rearrangements, some of which may have occurred prior to integration into cellular DNA (ROGLER and SUMMERS 1982, see Sect. 4.4).

The finding that Topo I mediates the integration of open circular DNA (WANG and ROGLER 1991) may have significance with regard to the maintenance of persistent infection. For example, if Topo I functions to divert open circular viral molecules from their normal processing to CCC DNA, the amount of viral DNA available as template for transcription could be lowered. Such a Topo I-mediated rearrangement-integration pathway would also lead to the production of defective, rearranged viral DNAs which would be incapable of replication. Molecules fitting these criteria, termed "novel forms," have been cloned from chronically infected woodchuck livers (ROGLER and SUMMERS 1982). Such a diversion pathway could also explain the gradual reduction of virus replication during persistent infection (RAIMONDO et al. 1988) and the build-up of viral integrants in the liver of long-term HBV carriers (SHAFRITZ et al. 1981). Viral CCC DNA itself may, of course, be another substrate for integration via the pathway suggested in Fig. 4.

4.3 Insertional Mutagenesis of Cellular Genes Involved in Growth Control

One of the primary goals of experiments designed to clone integrations over the past decade has been the identification of new genes involved in hepatic oncogenesis. The initial approach utilized cellular flanking sequences, adjacent to integrations from single integration tumors, as probes to search for common integration sites in tumors. The rationale for these experiments was that insertion of a viral transcriptional element adjacent to or near a cellular protooncogene would cause activation of the gene in a large subset (if not all) of HCC and therefor allow identification of protooncogenes involved in hepatocarcinogenesis. This approach was reasonable since the work of HAYWARD et al. (1981) had demonstrated that viral long terminal repeats (LTR) of avian leukosis virus activated the c-*myc* gene by a promoter insertion mechanism in bursal lymphomas of chickens.

Several significant cases of integration into cellular genes involved in growth control have also been discovered, and these may prove to be important in understanding the progression to HCC. In two cases, HBV integrations have identified new human genes with potential regulatory properties (DEJEAN et al. 1986, WANG et al. 1990), and in several other cases WHV integration occurred within and adjacent to the woodchuck c-*myc* and N-*myc* genes (MOROY et al. 1986; HSU et al. 1988; FOUREL et al. 1990). Whether the integrations in these genes were sufficient to initiate tumorigenesis or functioned to promote tumor progression awaits further experimentation. A brief description of these interesting cases is presented here.

One case of HBV integration into an intron of a human cyclin A gene has been reported (WANG et al. 1990). The integrated HBV DNA contained an HBV enhancer element, and it was speculated that this element could have been responsible for the increased steady state level of cyclin A mRNA observed in the tumor. Cyclins are characterized by their steady accumulation throughout interphase until the G2/m transition, followed by rapid disappearance at the onset of anaphase. Cyclins complex with and activate the serine-threonine kinase p34cdc2 during the G2/m phase transition (DRAETTA et al. 1989; DRAETTA and BEACH 1988) and are thought to be necessary for the initiation of mitosis and meiosis (MINSHULL et al. 1989). Their possible role in tumorigenesis could be through the control of phosphorylation of regulating proteins such as p60c-src or the retinoblastoma gene product, which is a potential inhibitor of cell proliferation (LUDLOW et al. 1989). In the specific case reported, the inserted HBV DNA contained a poly A addition site and a strong liver-specific promoter in addition to the enhancer element. These signals could have caused the production of amino-terminal and carboxyl-terminal truncated cyclin proteins of unknown (possibly deregulated) function.

In another instance HBV integration interrupted the coding region of the retinoic acid receptor B gene (DEJEAN et al. 1986; DETHE et al. 1987). In this case, integration occurred in an exon of the gene and created a chimeric viral-host open reading frame (ORF). The viral preS 1 promoter was present in the HBV sequences and served as the promoter for a viral-cellular cotranscript. Retinoic acid receptor proteins are nuclear DNA binding proteins which mediate the effects of retinoic acid (MCBURNEY et al. 1988). In some systems, retinoic acid is involved in cellular differentiation; however, its function in the precancerous liver or in HCC is unknown. Several cases are known in which mutant receptor genomes have either potentiated transformation or acted as transforming genes themselves.

The strongest evidence directly linking viral DNA integration to hepatocellular transformation is the discovery of frequent integration of WHV

DNA into an N-*myc* gene in woodchuck HCC's (FOUREL et al. 1990). The integrations occurred in exon 3 in the 3′ noncoding region of a unique woodchuck N-*myc*-2 cDNA gene and led to its transcriptional activation (FOUREL et al. 1990) WHV integration into the c-*myc* gene has also been observed in two woodchuck HCC's (MOROY et al. 1986; HSU et al. 1988). In one HCC the viral sequences, including the WHV enhancer element, integrated in exon 3, some 90 p downstream of the c-*myc* coding sequences. Preferential utilization of the c-*myc* p2 promoter occurred in the tumor and was associated with a 5–10-fold steady state increase in c-*myc* mRNA in the tumor. In the second tumor, a large (7 Kb), highly rearranged WHV DNA molecule integrated 600 bp upstream from the c-*myc* exon 1. This WHV DNA sequence contained a viral enhancer element, and analysis of c-*myc* RNA levels in this tumor revealed a 25-fold increase in c-*myc* RNA compared with the surrounding normal liver. In this case the c-*myc* P1 promoter was used fivefold more than the P2 promoter. Therefore, enhancer insertion appeared to be the most likely mechanism by which WHV caused c-*myc* RNA accumulation.

These examples, even though they have not yet defined a common mechanism of hepatic oncogenesis, have shown that viral integrations can be used as tools to identify cellular genes essential for cell growth and have directly implicated integrations in hepatocellular transformation in several specific cases. Future studies of the biological activities of these integrations and their associated genes in cell lines and transgenic mice are awaited to elucidate their role in hepatic oncogenesis further.

4.4 Integration as Agent of Genomic DNA Rearrangement

4.4.1 Microdeletions at Integration Sites

Nearly all the integrations cloned from precancerous liver contain viral DNA which is colinear with the viral genome but lacks some portion of the viral coding sequences due to deletion (ROGLER and SUMMERS 1984; YAGINUMA et al. 1987). Many integrations cloned from HCC also contain similarly deleted genomes (TOKINO et al. 1987; HINO et al. 1989). Based on the end specificity and polarity of viral sequences in integrations, four basic types of primary integrations have been proposed (SHIH et al. 1987) (Fig. 3). These include integrations with one virus-cell junction near the nick in the minus strand viral DNA extending either downstream through the core gene (type III) or the X-gene (type IV).

Type I and II integrations are the most common type cloned from chronically infected liver. Even though one end of a typical integration usually occurs in a specific region of the genome, the location of the opposite end is highly variable. Microdeletions of cellular DNA have also been found to occur at most HBV integration sites, and a description of some of the microdeletions which have occurred in association with HBV integrations is presented in Table 1. The deletions vary in length and nucleotide sequence. We analyzed 14 integration sites, and microdeletions occurred at 11 of them, macrodeletions at two, and a short direct duplication at one. Of the 26 junctions analyzed, 25 (96%) (two integrations had only one cellular junction sequenced) were positive for one or more preferred Topo I cleavage sites (underlined sequences in Table 1). This frequency is greater than that previously reported to be highly significant for illegitimate recombination in mammalian systems (KONOPKA 1988) and further supports the conclusion that Topo I plays a role in HBV integration in vivo.

4.4.2 Chromosome Rearrangements at Integration Sites

The majority of integrations cloned from fully developed HCC contain rearranged viral sequences (NAGAYA et al. 1987; SHAUL et al. 1984; OGSTON et al. 1982) and are often located at the site of major chromosomal abnormalities (TOKINO et al. 1987; MIZUSAWA et al. 1985; ROGLER et al. 1985; YAGINUMA et al. 1985; HINO et al. 1986). The DR sequences and the cohesive end region are preferred sites for viral–viral DNA rearrangements as they are for viral-cell junctions (TOKINO et al. 1987; HINO et al. 1986). This suggests that mechanisms similar to those responsible for primary integrations also function to generate viral-viral rearrangements. The most common rearrangements of viral sequences are inverted duplications accompanied by the deletion of viral sequences between the inverted segments. Direct duplications or short deletions are also common. Viral DNA rearrangements similar to these can be generated in vivo by Topo I, suggesting that this enzyme may be responsible for a portion of the rearrangements which occur in vivo (WANG et al. 1990).

Three major types of chromosome rearrangements have been reported in association with HBV integrations. These include deletion of large segments of the chromosome (ROGLER et al. 1985), inverted duplication of flanking sequences, and chromosomal translocations (MIZUSOWA et al. 1985; YAGINUMA 1985; TOKINO et al. 1987; HINO et al. 1989). In all these cases, HBV DNA integration appears to have mediated the rearrangement. Translocations associated with HBV integrations have occurred between chromosomes 17 and X, between chromosomes 9 and 5, and between chromosomes 17q22:18q11. Even though these results do not identify a unique chromosome translocation associated with HCC, they clearly demonstrate that integration mediates major chromosomal defects. The 17q22:18q11 translocation was accompanied by an additional deletion of chromosome 18q DNA at the translocation site (HINO et al. 1986).

Several integrations have been described which contain inverted HBV sequences flanked by inverted duplications of cellular DNA. In all of these cases the two viral-cell junctions are identical. Unfortunately, the boundary of the inverted cellular sequences with unrearranged cellular sequences was not determined. However, it is possible that these structures are generated by linking of the HBV DNA containing a hairpin loop to the end of a cellular DNA molecule followed by DNA replication to generate a double-stranded molecule with inverted repeats.

Table 1. Microdeletrons that have occurred in association with HBV integrations and preferred Topo I cleavage sites (underlined)

Integration Number	Clone	Topo I Sites	5′	R	Deletion	R	3′	Topo I Sites	Reference
1	λDT1	4	CCATC GGTAG	TC AG	cacttcc gtgaagg	TCC AGG	CCCTC GGGAG	3	DEJEAN et al. (1986)
2	HC1701-1	4	CAAAA GTTTT	T A	gagattggtggaggaaa ctctaaccacctccttt		AATTT TTAAA	2	YAGINUMA et al. (1987)
3	c3	1	TTTAG AAATC	T A	atatc tatag	TTAAA AATTT	TTTAG AAATG	4	HINO et al. (1989)
4	C29	2	CACGC GTGCG	TG AC	taa att	T A	CCCAG GGGTC	2	NAGAYA et al. (1987)
5	p2	0	GCGCG CGCGC	G C	tggcgcgcgcctgta accgcgcgcggacat		GTCCC CAGGG	1	NAGAYA et al. (1987)
6	pY	2	TCCCC AGGGG		aaattcctcttg tttaaggagaac	TAA ATT	TTGAT AACTA	4	NAGAYA et al. (1987)
7	W93HCC	2	ACAAC TGTTG		ctc gag		AAGAA TTCTT	2	HSU et al. (1988)
8	PLCPRF5-26	1	TTTTT AAAAA	AACG TTGC	tgtagccgggcgtccat acatcggcccgcaggta		GATAC CTATG	3	BERGER and SHAUL (1987)
9	HuH2-2	3	CTGAG GACTC	GGCCTTTGTC CCGGAAACAG			CCAAT GGTTA	2	YAGINUMA et al. (1985)
10	PLCPRF5-14	1	TCACA AGTGT		acagg. . . . . tgtcc.				
					. . . . .gggga . . . . .cccct	AGG TCC	CCAGC GGTCG	1	BERGER and SHAUL (1987)

11	H2-7	3	TCTTC TG AGAAG AC	agctt ..... tcgaa.....			
				......tgta ......acat	CAACT ATTTT GTTGA TAAAA	2	Takada et al. (1990)
12	HCY-23	3	GCACT CTAGA CGTGA GATCT	tttaa ..... aaatt.....			
				.....accac .....tggtg	AG GCCAT TC CGGTA	1	Takada et al. (1990)
13	HT14NCP	2	CTAGA CA GATCT GT	gaagc..... cttcg......			Shih et al. (1987)
14	W74HCC	2	GACTG AA CTGAC TT	ctgca..... gacgt.....			Hsu et al. (1988)

Analysis for	Total no. analyzed	No. of positive	Percentage positive
Junctions positive for Topo I site(s)	26	25	96
Junctions positive for multiple Topo I sites	26	19	73

Deletions of large segments of cellular DNA have also been reported in association with HBV integration. One interesting case was that in which HBV integration mediated a deletion of more than 13 kb of cellular DNA at chromosome locus 11p13-14 (ROGLER et al. 1985). Since the Wilms' tumor suppressor gene is located at chromosome locus 11p13 (FRANCKE and RICCARDI 1979), this raised the question as to whether the HBV integration caused deletion of one copy of the Wilms' tumor suppressor gene. As the Wilms' tumor suppressor gene has not yet been cloned, the actual loss of the gene in the tumor awaits further analysis. The possible role of tumor suppressor gene inactivation in HCC is discussed in Sect. 5.5.

4.5 Hepadnavirus Proteins and Hepatocarcinogenesis

Until recently, no biological function had been described for the protein encoded by the X open reading frame of mammalian hepadnaviruses. Due to its nucleotide composition and position in the genome, relative to the position of oncogenes in retroviruses, investigators have suggested that the X-gene was originally a cellular gene (MILLER and ROBINSON 1986). The X-genes of HBV, WHV, and GSHV have since been shown to encode proteins which transactivate transcription from a wide array of target promoters (TWU and SCHLOEMER 1987; SPANDAU and LEE 1988; ZAHM et al. 1988; TWU and ROBINSON 1989; COLGROVE et al. 1989). Transactivation of both hepadnavirus and heterologous promotors has been observed when the X protein was expressed in its native state during productive infection (COLGROVE et al. 1989). Northern blot analysis of reporter gene RNA levels and transcription run on analyses indicate that the *trans*-activation occurs at the level of primary transcription.

Computer analysis of X protein sequences predict an X polypeptide which is a small, soluble, intracellular protein without a signal sequence or other obvious identifying sequence motifs (COLGROVE et al. 1989). Its transactivation activity varies to a great degree in different cell lines and is usually the greatest in liver-derived cell lines (TWU and ROBINSON 1989). However, large differences in trans-activation occur among different hepatoma cell lines using the same target promoter (TWU and ROBINSON 1989). These results suggest that X protein trans-activation requires the participation of other cellular factors, and a recent report has shown that cellular transcription factors Ap1, Ap2, and Sp1 can mediate some X protein transactivation effects (SETO et al. 1990). Indeed, X protein has recently been shown to be a novel protein kinase (T.T. WU and W.S. ROBINSON 1990).

Since the 3′ end of the X-gene lies within the cohesive overlap, integration of hepadnavirus DNA frequently occurs within the X ORF and results in truncation of the X protein coding sequence. In several of these cases, the X proteins expressed are fusion proteins with cellular domains. These X-cellular fusion proteins, which are expressed in HCC, have been shown to maintain their trans-activation potential (WOLLERSHEIM et al. 1988).

Human T-cell leukemia virus (HTLV-1) and bovine leukosis virus (BLV) also contain transactivating factors and appear to participate in oncogenesis in a manner similar to hepadnaviruses. Neither HTLV-1 nor BLV encode their own oncogenes or integrate their proviral DNA in a site-specific manner in tumors (WONG-STAAL and GALLO 1985). However, both encode proteins responsible for transcriptional activation of viral gene expression (MARUYAMA et al. 1987). In addition, the HTLV-1 transactivator gene activates transcription of the interleukin-2 receptor gene (GREENE et al. 1986), suggesting that it plays a role in stimulating lymphoid cell proliferation. In adenovirus, the transcriptional trans-activator protein coded by the E1a region has been shown to potentiate the transforming effects of viral and cellular oncogenes (RULEY 1983).

The first evidence directly linking X-gene expression with pathologic changes in the liver has recently been obtained through the use of HBV-X-gene transgenic mice. Two lines of these mice were developed by Dr. GILBERT JAY by injecting mouse oocytes with a HBV DNA fragment comprising the HBV enhancer, X promoter, X ORF, and HBV polyadenylation signal. X-gene expression in all three lines caused a lesion which began as small focal areas of hepatocytes described as clear cells, due to the presence of clear vacuoles which contained large amounts of glycogen. These lesions eventually progressed to involve the whole liver by the age of 1 year, and eventually clear cell adenomas developed (Dr. GILBERT JAY, personal communication). These findings which directly link the HBV-X-gene with pathological changes which occur in the early stages of hepatocarcinogenesis are in contrast to data from other laboratories in which HBV- X-gene transgenic mice did not develop a pathological reaction (LEE et al. 1990). They also demonstrate that the X protein is not a complete oncogene and that additional genetic change is required for continued progression to HCC.

A second transcriptional transactivator has most recently been identified in the preS2/S region of an integrated HBV DNA which was cloned from a hepatocellular tumor (KEHULE et al. 1990). The transactivation activity was produced by a 3′-truncated preS2/S region fused to cellular DNA which encoded a 102 aa viral-cellular fusion protein. A frame shift mutation within the preS2 ORF of the fusion protein completely eliminated the transactivation activity, indicating that the protein itself was responsible for the effects on transcription. An artificially 3′-truncated preS2/S gene from a wild-type HBV genome also exhibited transactivation activity while the intact HBV preS2/S gene was not active as a transactivator. Thus, the cellular sequences were not necessary for the transactivation effect. Studies of the steady state level of reporter gene mRNA showed that the transactivation effect occurred at the transcriptional level.

Since the HuH-4 cell line, from which the original pS2/S transactivator was cloned, strongly expressed the c-*myc* gene, an experiment was conducted to determine whether the pS2/S transactivator could transactivate c-*myc* promoters linked to a reporter gene. The experiments demonstrated that it was capable of transactivating the P2 but not the P_1 promoter of the c-*myc* gene. Interestingly, truncated preS2/S genes are present in many of the integrations which have been cloned from HCC (Shaul et al. 1984; Ogston et al. 1982), and a second integration containing a similar 3′-truncated preS2/S sequence also exhibited transactivation activity. These findings support the idea that accidental 3′-truncations of integrated HBV preS2/S genes may frequently contribute to hepatocellular transformation, possibly through the activation of cellular protooncogenes.

5 Studies on the Activities of Protooncogenes, Growth Factors, Tumor Suppressor Genes, Immune Surveillance, Cell–Cell Interactions, Cellular Differentiation, etc. During Carcinogenesis

5.1 General Considerations Regarding the Complexity of Experimental Carcinogenesis

The title of this section was purposely composed to emphasize the complexity of the subject of hepatocarcinogenesis. The biological effects of specific protooncogenes, growth factors, and tumor suppressor genes are undoubtedly affected by factors which vary with the cell type and phase of the cell cycle. Although the predominant cell type in the liver is the hepatocyte, others such as endothelial cells, Kupffer's cells, and inflammatory cells (especially in the case of viral infection) comprise a significant portion (~10%–20%) of the liver mass. In various precancerous states, significant variations in cell populations occur. For example, oval cell proliferation precedes and leads to the formation of precancerous nodules, and endothelial cell proliferation can lead to fibrosis and cirrhosis. Fibroblasts produce a family of growth factors (fibroblast growth factors, FGFs) which induce angiogenesis, and these factors are most certainly necessary for the vascular development of tumors (Burgess and Maciag 1989). Their effects on hepatocytes and tumor growth are unknown. Furthermore, inflammatory cells produce a series of interleukins, and platelets produce hepatocyte growth factor (HGF), which is a potent stimulator of hepatocyte growth (Nakamura et al. 1989).

Our understanding of the specific cellular and genetic controls which are altered in initiated hepatocytes in precancerous nodules or in transformed hepatocytes is very incomplete. In fact, there is currently no consensus as to the specific protooncogene activities associated with any stage of hepatocarcinogenesis in vivo. Intense study has shown that protooncogene expression can be controlled by gene amplification, gene rearrangement, point mutations, altered transcription (up or down and alternate splicing), and translational and posttranslational mechanisms (Friend et al. 1989). At the present time, no single protooncogene or growth factor has been studied thoroughly enough to rule out or establish its universal participation in hepatic oncogenesis.

Tumor cells typically carry multiple genetic changes that act synergistically to produce the full range of neoplastic traits (FRIEND et al. 1989). The requirement of multiple oncogenes is quite consistent with multistep carcinogenesis in the liver. Each step through which a normal cell proceeds on the route to malignancy may be demarcated by the acquisition of one mutant gene or another. Mating experiments with transgenic mice containing liver-targeted oncogenes have clearly shown that oncogenes function synergistically together to induce HCC (SANDGREN et al. 1989). The problem remains which and how many oncogenes are activated or suppressed during naturally occurring HCC. Also, do the activities of oncogenes which participate in the early stages of oncogenesis continue to be required after the cells have progressed further toward malignancy? In some cases, specific oncogenes involved in the premalignant phases of carcinogenesis have been proposed, as is the case for the *ras* oncogenes in premalignant papillomas on mouse skin (BALMAN and PRAGNELL 1983) and the evolution of benign colon polyps into frank carcinomas in the colon (FORRESTER et al. 1987). In the absence of a consensus as to oncogenes and growth factors involved at different stages in hepatic oncogenesis, it is only possible at present to cite examples in which specific oncogenes and growth factors have been implicated in the oncogenic process in liver. A complete understanding of the complex interactions involved in hepatocellular transformation awaits further investigations.

5.2 The *myc* Family of Protooncogenes

One study of c-*myc* expression during liver regeneration, after partial hepatectomy, has shown that c-*myc* mRNA accumulates in a transient manner within the 1st h after surgery followed by a second and also transient activation 8 h later (THOMPSON et al. 1986). Another study showed that c-*myc* mRNA remained elevated during the entire regeneration response (SOBCZAK et al. 1989). Transcription of c-*myc* was studied in a set of related liver epithelial lines (hepatocyte-like oval cells) which exhibited untransformed, partially transformed, or fully transformed phenotypes (BRAUN et al. 1989). The c-*myc* transcription was cell cycle regulated in the untransformed and partially transformed cell lines, similar to its regulation during liver regeneration. However, after continuous passage in vitro the cells acquired the capacity to form tumors, and at this point, expression of c-*myc* was found to be constitutive (BRAUN et al. 1989). Constitutive expression of c-*myc* also occurs in the HepG2 hepatoma cell line, and the accumulation of c-*myc* mRNA is not a consequence of the increased stability of the message (HUBER and THORGEIRSON 1987). Thus, deregulation of c-*myc* transcription may be a general characteristic of hepatocellular tumor cells. As stated earlier, WHV integration also appears to play a direct role in *myc* gene activation in virus carriers (Sect. 4.3).

Studies which have utilized diet as the carcinogenic regimen in rats have supported a role for the c-*myc* gene in precancerous as well as carcinogenic changes in the liver (CHANDAR et al. 1989). When rats were fed a choline-devoid (CD) diet for 12 months followed by a choline-supplemented diet, 73% of the animals developed moderately well-differentiated HCC of trabecular adenomatous or mixed phenotypes. Amplification of the c-*myc* gene was found in 13 of the 13 tumors. The amplification ranged from 2- to 70-fold and was accompanied by increases in c-*myc* gene expression. In addition, gene amplification was also apparent in nontumorous regions of tumor-bearing livers. The authors proposed that gene amplification occurs at many loci (endogenous retroviral loci were also rearranged) during the CD diet and that cells containing c-*myc* amplifications have a selective advantage. A two- to five fold increase in the c-*myc* steady state mRNA level occurred in both livers and tumors of tissues with c-*myc* amplifications. This modest increase appears to provide enough additional c-*myc* expression to deregulate cell proliferation.

The involvement of c-*myc* in hepatocarcinogenesis is also supported by studies with transgenic mice in which a c-*myc* gene was placed under the control of an albumin promotor (SANDGREN 1989). These animals develop mildly hyperplastic lesions in the liver which progress to form benign adenomas. Thus, c-*myc* initiates limited cell regeneration in the context of differentiated hepatocytes in an intact liver. When c-*myc* transgenic mice were mated with transgenic mice containing an activated c-H-*ras* gene, the transgenic littermates which contained both the activated c-*myc* and c-H-*ras* genes developed HCC at a very rapid rate compared with littermates containing either oncogene alone (SANDGREN 1989). However, the HCC appeared as clonal growths protruding from the nontumorous, hyperplastic liver. Therefore, even in these mice additional genetic events were clearly required for hepatocyte transformation, and identification of these additional genetic changes is one of the greatest challenges in tumor biology today.

5.3 *ras*, *raf*, and Other Protooncogenes Activated During Hepatocarcinogenesis

The utilization of standard transfection assays to search for activated cellular protooncogenes in HCC have yielded mostly negative results. This is especially true for studies which have utilized fibroblast cell lines to assay for protooncogenes in the *ras* family. Some reports have, however, detected oncogenes in HBV-associated HCC (OCHIYDA et al. 1986; TAKADA and KOIKE 1989). A problem with this approach has been that the assays are limited to the detection of dominantly transforming genes which can function in both fibroblasts in vitro and hepatocytes in vivo. Recently, a liver epithelial cell line, with oval cell markers, was transformed by transfection with the EJ *ras* oncogene, and the transformed cells gave rise to HCC when injected into nude mice (BRAUN et al. 1987). The nude mouse tumor exhibited a wide spectrum of phenotypes from anaplastic, poorly differentiated HCC to moderately well differentiated HCC.

Such differences would be expected to occur from transfection of a pluripotent oval cell line, and further work in this area may be of value.

The complexity of this issue of hepatocyte transformation is perhaps most easily seen in studies involving chemical carcinogenesis, which have clearly shown that tumor formation varies with and is dependent on the species and strain of rodent employed. Thus, host variation may significantly affect the progress to HCC in hepadnavirus infections. B6C3F1 mice, for instance, have a high spontaneous rate of HCC which is thought to be due to the loss of tumor suppressor genes (as yet unidentified) in these strains. Consistent with this, investigators have been able to detect oncogenes in HCC DNA of B6C3F1 mice. For example, some reports have detected oncogenes in up to 80% of spontaneously occurring HCC (REYNOLDS et al. 1986, 1987). The majority of the oncogenes activated were c-H-*ras* genes with a lesion in codon 61. Activation of the H-*ras* gene was dominant compared with the N-*ras* (0/13) and K-*ras* (2/13) genes. A broad spectrum of activating point mutations in the H-*ras* gene was observed. Interestingly, foci induced by mutations in H-*ras* codons 13 and 117 were slower to appear and less morphologically transformed than foci induced by codon 12 or codon 61 H-*ras* mutants. This could in part account for why these mutations were not previously detected in HCC. Novel, weakly activating point mutations in *raf* genes and four non-*ras* transforming genes were also observed in the experiments (REYNOLDS et al. 1987). Activated oncogenes were also detected in 30% of hepatocellular adenomas in B6C3F1 mice, suggesting that oncogene activation may confer a selective advantage on premalignant hepatocytes (REYNOLDS et al. 1986). Forty percent of the mouse hepatocellular adenomas contained some morphologic features of malignancy, indicating that they are precursors of HCC. Thus, mutations in the *ras* family of protooncogenes (preferably H-*ras*) are capable of expressing themselves in high frequency in the B6C3F1 genetic background and may have a role in precancerous or cancerous changes in some but not all hepadnavirus patients.

5.4 Possible Roles of Paracrine and Autocrine Growth Factors such as TGF-α, TGF-β, HGF, and IGF-II in Hepatocarcinogenesis

A complex array of growth factors and their receptors have been identified which affect the growth and differentiation of numerous cell types. In most cases, these growth factors act through autocrine or paracrine mechanisms to regulate the growth of target cells. Since growth factor- and/or growth factor receptor-induced polyclonal proliferation of cancer precursor cells are important aspects of the etiology of several types of tumors, it is very likely that endogenous liver growth factors also play significant roles in the etiology of HCC.

Current knowledge of the activity of endogenous liver growth factors during liver development or progression to HCC is very fragmentary. The understanding which has been gained has come primarily through studies of growth factor expression during liver regeneration and through analysis of growth factor expression in precancerous liver and HCC. Four endogenous liver growth factors which vary significantly in their activity under different experimental regimens are tumor growth factor alpha (TGF-α), tumor growth factor beta (TGF-B), hepatocyte growth factor (HGF), and insulin-like growth factor II (IGF-II). A brief discussion of some of the evidence linking them to liver cell and tumor growth are presented.

5.4.1 Tumor Growth Factor Alpha

TGF-α is a 50-aa peptide closely related to epidermal factor (EGF). It binds to the same receptor as EGF on hepatocyte plasma membranes and in general has similar biological effects. Increases in TGF-α mRNA in hepatocytes coincide with the peak of hepatocyte DNA synthesis between 24 and 72 h after partial hepatectomy (MEAD and FAUSTO 1989). Addition of TGF-α 20 ng/ml to hepatocyte cultures stimulates approximately a 13-fold increase in DNA synthesis, and this increase is inhibited when TGF-β is added at 100 pg/ml. TGF-α protein is doubled in regenerating liver 24 h after partial hepatectomy, and cultured hepatocytes can secrete TGF-α.

EGF receptors are present on hepatocyte membranes after partial hepatectomy, and some reports have detected modest reductions in receptor mRNAs and proteins. It is difficult to implicate EGF as the ligand responsible for down-regulation of its own receptor during liver regeneration since it is not made in the liver, and its serum concentration does not change after partial hepatectomy (RALL et al. 1985). In contrast, the above results implicate TGF-α as the physiological ligand that participates in an autocrine loop with EGF receptors. TGF-α does not appear to initiate the regenerative response but comes into play later, probably at a time when hepatocytes have already made the transition from G_0 to G_1 and have entered the cell cycle (MEAD and FAUSTO 1989). In spite of its clear role in liver regeneration, no studies have yet shown its expression to be altered (up or down) in either precancerous livers or HCC.

The mature form of TGF-α is cleaved from an integral membrane precursor. The absence of an appropriate protease or protease cleavage site would ensure that the growth factor remains anchored to the plasma membrane. A recent study demonstrated that TGF-α precursor molecules, anchored in the cell surface, can bind to the EGF receptor on adjacent cells, leading to signal transduction (WONG et al. 1989). Such molecules continued to stimulate activity of the EGF receptor kinase on adjacent cells (WONG et al. 1989). It was proposed that membrane-anchored growth factors may lead to prolonged biological responses relative to those of secreted growth factors. Whether such activities are involved in liver cell proliferation and hepatocellular carcinoma is unknown; however, these results point out the limitations of conclusions based solely on mRNA steady state accumulation levels.

5.4.2 Tumor Growth Factor Beta

Partial hepatectomy and the consequent rise and fall of liver regeneration also provide an opportunity to study factors that terminate hepatocyte replication. TGF-β is a potent inhibitor of DNA synthesis in primary monolayer cultures of adult rat hepatocytes (BRAUN et al. 1986). TGF-β $5 \times 10^{-12} M$ also induced a 50% inhibition of EGF-mediated DNA synthesis in primary hepatocytes (BRAUN et al. 1986). Specific saturable receptors for TGF-β were found on normal rat hepatocytes, but specific binding could not be detected on hepatocytes isolated from 24-h regenerating liver (BRAUN et al. 1986). These studies in conjunction with others which show that TGF-β is produced by endothelial cells after partial hepatectomy (THOMPSON et al. 1986) suggest that TGF-β inhibits hepatocyte growth by a paracrine mechanism in the late stages of the partial hepatectomy response.

TGF-β also exerts vastly different effects on the proliferation of normal rat liver epithelial cells versus transformed liver cells. TGF-β inhibited colony formation (83%) and DNA synthesis of immortalized rat liver epithelial cells and primary hepatocytes at levels as low as 20 pg/ml (MCMAHON et al. 1986). Transformation of the immortalized liver epithelial cells by treatment with aflatoxin in vitro was associated with a loss of sensitivity of the cells to inhibition by TGF-β. TGF-β inhibition of normal liver epithelial cells caused morphological changes characterized by cytoplasmic hypertrophy and formation of abundant lysosomal derivatives, resembling lipofuscin. One possibility in this context is that TGF-β promotes differentiated functions in these cells.

In HCC of WHV-carrier woodchucks, TGF-β mRNA accumulation occurs only in isolated megakaryocytes (YANG and ROGLER, unpublished data). In SV40-induced HCC in transgenic mice, fibroblast proliferation surrounds and fills up regions of necrosis. TGB-β mRNA accumulation occurs in the regions of fibroblast growth and not in the HCC of these animals (P. SCHIRMACHER and C.E. ROGLER, unpublished data). In these cases, hepatocellular tumor cells may have lost their sensitivity to TGF-β, as occurred in the in vitro transformed liver epithelial cells described above.

5.4.3 Hepatocyte Growth Factor

It has long been known that a serum growth factor is present in rats after partial hepatectomy and other types of chemically induced liver damage. Recently, HGF was purified to homogeneity, and the purified protein was used to clone a full length cDNA (NAKAMURA et al. 1989). HGF is produced in and released from platelets which accumulate in the liver as a result of various types of liver injury. Purified preparations of HGF are potent stimulators of DNA synthesis in primary hepatocytes (NAKAMURA et al. 1986). Studies of the mechanism of this paracrine induction of hepatocyte growth are in their infancy. The initial report of the cloning of HGF cDNA stated but did not present proof that HGF is also produced by hepatocytes. In any event, platelets accumulate in the liver of hepadnavirus carriers, and it is therefore quite likely that this growth factor plays a significant role in the etiology of HCC in hepadnavirus carriers.

5.4.4 Insulin-Like Growth Factor II

IGF-II (originally called somatomedin A) was originally identified as a group of polypeptides present in rat serum and the conditioned medium of a buffalo rat liver epithelial cell line, called BRL-3A (Dulak and Temin 1973). In addition to inducing insulin-like effects in various bioassays, IGF-II is a potent mitogenic growth factor for chicken embryo fibroblasts in vitro (Dulak and Temin 1973). It also increases DNA synthesis in primary hepatocyte cultures (Rechler and Nissley 1985). The single IGF-II gene is expressed at very high levels in fetal liver, which is the main source of IGF-II in the fetus (Soares et al. 1985). However, the transcription rate declines after birth and is almost undetectable in normal adult liver. Reactivation of IGF-II transcription has been observed in precancerous and HCC tissues in rats exposed to carcinogenic regimens, in WHV-carrier woodchucks, and in HCC from human HBV carriers (Fu et al. 1988; Cariani et al. 1988; Uneo et al. 1988).

While many examples of increased IGF-II transcription have been described, the actual function of IGF-II in vivo has remained elusive. IGF-II biology is complicated by the presence of IGF-II binding proteins in serum which appear to lower its biological activity (Romanus et al. 1986). The identification of the IGF-II receptor as the mannose-6-phosphate receptor which is abundant in hepatocytes as well as many other cells has further complicated understanding of its biology (Morgan et al. 1987). Currently, most but not all evidence supports the notion that IGF-II functions through its moderately strong cross-reactivity with IGF-I receptors (Rechler and Nissley 1985), and the function of the strong binding to the mannose-6-phosphate receptor is unknown.

Both the rat and human IGF-II genes have three promoters which are differentially activated in various tissues (Ueno et al. 1988). The PII promoter of rat IGF-II is highly active in fetal tissue and inactive in adult tissue whereas the PI promoter is most active in adult liver (albeit at a low level). IGF-II primary transcripts are differentially spliced (Soares et al. 1986) and exhibit different translation efficiencies in vitro (Rechler and Nissley 1985). The fetal IGF-II mRNA transcribed from the PII promoter appears to be the most efficiently translated IGF-II mRNA (Rechler and Nissley 1985).

Accumulation of IGF-II mRNA corresponding in size to fetal IGF-II mRNA has been observed in precancerous liver and HCC of WHV-carrier woodchucks and HBV carriers (Fu et al. 1988; Cariani et al. 1988). In situ hybridization studies have also provided circumstantial evidence that IGF-II plays a role in liver growth and carcinogenesis in hepadnavirus carriers. In persistently infected woodchuck livers, precancerous nodules have been identified. Many of the nodules exhibited a slight elevation in IGF-II mRNA accumulation, uniformly throughout the nodule, and in addition focal groups of cells containing very high levels of IGF-II mRNA were observed within some of the nodules (Yang and Rogler

1991). In conjunction with this, almost all HCC arising in woodchucks contain regions of highly elevated IGF-II mRNA accumulation. In many instances high IGF-II mRNA is associated primarily with the extreme growing edge of tumors, as judged by a high number of hepatocytes in S phase. These data support the notion that cells within precancerous nodules which express high levels of IGF-II are at a selective advantage for tumor progression and that IGF-II promotes the growth of malignant HCC.

5.5 Tumor Suppressor Genes

It is generally accepted that the inactivation of tumor suppressor genes is one of the prerequisites for malignant transformation in many types of cells (FRIEND et al. 1989). The involvement of tumor suppressor genes in HCC was first suggested by the occurrence of hepatoblastoma in patients with Beckwith–Wiedemann syndrome (KOUFOS et al. 1985). This syndrome is associated with duplications and rearrangements of genes on chromosome 11 p. The Wilms' tumor suppressor gene has also been localize to chromosome 11 p 13. Therefore, the identification of an HBV integration at chromosome locus 11p13 and an associated deletion of at least 13 kb of cellular DNA suggested that HBV integration could have caused the loss of one copy of a tumor suppressor gene in that tumor (ROGLER et al. 1985).

HBV integrations have been mapped to 11 chromosomes, showing that integration occurs widely over many chromosomes (NAGAYA et al. 1987; TOKINO et al. 1987). However, five cases the HBV integration in chromosome 11 have been reported, and this figure is higher than expected for random integration (NAGAYA et al. 1987). Another study used restriction fragment length polymorphisms (RFLPs) to study chromosome alterations in HCC from HBV carriers. Panels of genes from chromosome 11p and 13q plus individual genes from 19 other chromosomes were used to study chromosome abnormalities in the tumors. The study revealed a high frequency of loss of alleles from chromosomes 11p (45%) and 13q (50%) in HCC DNA (WANG and ROGLER 1988). In addition to the Wilms' locus on 11p13, the retinoblastoma tumor suppressor gene is located on chromosome 13q. Therefore, the selective loss of alleles in chromosomes 11p and 13q supports the notion that the loss of tumor suppressor genes is responsible for part of the mechanism of hepatocarcinogenesis.

Transgenic mice which express SV40 T antigen in the liver develop HCC after a period of prolonged liver hyperplasia (HELD et al. 1989). Cell culture studies have also shown that SV40 T Ag is capable of immortalizing hepatocytes, although the mechanism is unknown. However, in other transgenic mice and in cell cultures, SV40 T Ag has been shown to complex with the retinoblastoma (Rb) protein from chromosome locus 13q (DECAPRIO et al. 1988; LUDLOW et al. 1989). The biological consequences of SV40 T Ag binding to the Rb protein are not fully understood; however, some studies indicate that T Ag binding to the Rb protein affects the latter's phosphorylation, which in turn alters its functional activity in cell cycle regulation (DeCaprio et al. 1988). Thus, a major function of SV40 T Ag is likely to involve alteration of the activity of tumor suppressor genes.

When SV40 T Ag expression is combined with the overexpression of a cellular protooncogene, such as the H-*ras* gene in transgenic mice, HCC develops much more rapidly in the mice with both oncogenes than in those with each oncogene (Sandgren et al. 1989). Similar to the c-*myc* plus H-*ras* mice, the tumors arising in the SV40 T Ag plus H-*ras* mice do not involve the whole liver and, therefore, arise through the selection of cells which have undergone additional genetic or epigenetic alterations. In addition to the presence of two activated oncogenes, almost all the HCC appearing in SV40 T Ag transgenic mouse liver expresses high levels of IGF-II (P. Schirmacher and C.E. Rogler, unpublished data), raising the possibility that derepression of IGF-II expression is one of the additional events required for progression to HCC. Interestingly, in humans, the IGF-II gene is located at chromosome 11p15 and the Wilms' tumor suppressor gene is located at chromosome 11p13. This suggests that loss of a suppressor gene at 11p13 could lead to derepression of a growth factor gene, such as IGF-II, at a distant site on the same chromosome. These studies are just beginning to uncover the complex interactions which mediate the activities of different classes of oncogenes.

References

Balman A, Pragnell IB (1983) Mouse skin carcinomas induced *in vivo* by chemical carcinogens have a transforming Harvey-ras oncogene. Nature 303: 72–74

Berger I, Shaul Y (1987) Integration of hepatitis B virus: Analysis of unoccupied sites. J Virol 61: 1180–1186

Beasley RP, Lin CC, Hwang LY, Chien CS (1981) Hepatocellular carcinoma and hepatitis B virus. A perspective study of 22,707 men in Taiwan. Lancet 2: 1129–1133

Braun A, Carr BI, Hayashi I, Branum EL, Moses HL (1986) Inhibition of DNA synthesis in rat hepatocytes by platelet-derived type B transforming growth factor type B. Proc Natl Acad Sci USA 85: 1539–1543

Braun L, Goyette M, Yaswen P, Thompson NL, Fausto N (1987) Growth in culture and turmorigenicity after transfection with the ras oncogene of liver epithelial cells from carcinogen treated rats. Cancer Res 47: 4116–4124

Braun L, Mikumo R, Fausto N (1989) Production of hepatocellular carcinoma by oval cells: cell cycle expression of c-myc and p53 at different stages of oval cell transformation. Cancer Res 49: 1554–1561

Brechot CM, Pourcel C, Louise A, Rain B, Tiollais P (1980) Presence of integrated hepatitis B virus DNA sequences in cellular DNA in human hepatocellular carcinoma. Nature 286: 533–535

Brechot CM, Hadchouel M, scotto J, Fonck M, Potet F, Vyas GN, Tiollais P (1981) State of hepatitis B virus DNA in hepatocytes of patients with hepatitis B surface antigen-positive and- negative liver disease. Proc Natl Acad Sci USA 78: 3906–3910

Brill SJ, DiNardo S, Voelkel-Meiman K, Sternglantz R (1987) Need for DNA topoisomerase I activity as a swivel for DNA replication and for transcription of ribosomal RNA. Nature 326: 414–416

Burgess WH, Maciag T (1989) The heparin-binding (fibroblast) growth factor family of proteins. Annu Rev Biochem 58: 575–606

Cariani E, Lasserre C, Seurin D, Hamelin B, Kemeny F, Franco D, Czech MP, Ullrich A, Brechot C (1988) Differential expression of insulin-like growth factor II mRNA in human primary liver cancers, benign liver tumors and liver tumors and liver cirrhosis. Cancer Res 48: 6844–6849

Chakraborty PR, Ruiz-Opazo N, Shouval D, Shafritz DA (1980) Identification of integrated hepatitis B virus DNA and expression of viral RNA in and HBsAg-producing human hepatocellular carcinoma cell line. Nature 286: 531–533
Champoux JJ (1977) strand breakage by the DNA untwisting enzyme results in the covalent attachment of the enzyme to DNA. Proc Natl Acad Sci USA 74: 3800–3804
Champoux JJ (1978) Mechanism of the reaction catalyzed by the DNA untwisting enzyme: attachment of the enzyme to the 3′-terminus of the nicked DNA. J Mol Biol 118: 441–446
Champoux JJ (1988) Topoisomerase I is preferentially associated with isolated replicating simian virus 40 molecules after treatment of infected cells with camptothecin. J Virol 62: 3675–3683
Champoux JJ, Bullock PA (1988) A possible role for the eucaryotic type I topoisomerase intellegitimate recombination. In: Kucherlapati R, Smith G (eds) Genetic recombination, American Society for Microbiology, Washington, pp 655–656
Chandar N, Lombardi B, Locker J (1989) C-myc gene amplification during hepatocarcinogenesis by a choline-devoid diet. Proc Natl Acad Sci USA 86: 2703–2707
Chen DS, Hoyer BH, Nelson J, Pourcel RH, Gerin JL (1982) Detection and properties of hepatitis B viral DNA in liver tissues from patients with hepatocellular carcinoma. Hepatology 2: 42S–46S
Chisari FV, Ferrari C, Mondelli MU (1989a) Hepatitis B virus structure and biology-mini review. Microbiol Pathog 6: 311–325
Chisari FV, Filippi P, Buras J, McLachlan A, Popper H, Pinkert CA, Palmiter RD, and Brinster RL (1987) Structural and pathological effects of synthesis of hepatitis B virus large envelope polypeptide in transgenic mice. Proc Natl Acad Sci USA 84: 6909–6913
Chisari FV Klopchin K, Moriyama T, Pasquinelli C, Dunsford HA, Sell S, Pinkert CA, Brinster RL, Palmiter RD (1989b) Molecular pathogenesis of hepatocellular carcinoma in hepatitis B virus transgenic mice. Cell 59: 1145–1156
Colgrove R, Simon G, Ganem D (1989) Transcriptional activation of homologous and heterologous genes by the hepatitis B virus gene product in cells permissive for viral replication. J Virol 63: 4019–4026
DeCaprio J, Ludlow JW, Figge J, Shew JY, Huang CM, Lee WH, Marsilio E, Paucha E, Livingston DM (1988) SV40 large tumor antigen forms a specific complex with the product of the retinoblastoma susceptibility gene. Cell 54: 275–283
Dejean A, Sonigo P, Wain-Hobson S, Tiollais P (1984) Specific hepatitis B virus integration in hepatocelular carcinoma DNA through a viral 11-base-pair direct repeat. Proc Natl Acad Sci USA 81: 5350–5354
Dejean A, Bougueleret L, Grzeschik K-H, Tiollais P (1986) Hepatitis B virus DNA integration in a sequence homologous tov-erb-A and steroid receptor genes in a hepatocellular carcinoma. Nature 322: 70–72
DeThe H, Marchio A, Tiollais P (1987) A novel steroid thyroid hormone receptor-related gene in appropriately expressed in human hepatocellular carcinoma. Nature 330: 667–670
Draetta G, Beach D (1988) Activation of cdc2 protein kinase during mitosis in human cells: Cell cycle-dependent phosphorylation and subunit rearrangement. Cell 54: 17–26
Draetta G, Luca F, Westendorf J, Brizuela L, Ruderman J, Beach D (1989) cdc2 protein kinase is complexed with both cyclin a and b: evidence for proteolytic inactivation of mpf. Cell 56: 829–838
Dulak NC, Temin HM (1973) Multiplication-stimulating activity of chicken embryo fibroblasts from rat liver cell conditioned medium: a family of small polypeptides. J Cell Physiol 82: 161–170
Dunsford HA, Sell S (1989) Production of monoclonal antibodies to preneoplastic liver cell populations induced by chemical carcinogens in rats and to transplantable Morris hepatoma. Cancer Res 49: 4984–4900
Dunsford HA, Karnasuta C, Hunt JM, Sell S (1989) Different lineages of chemically induced hepatocellular carcinoma in rats defined by monoclonal antibodies. Cancer Res 49: 4894–4900
Edman JC, Gray P, Valenzuela P, Rall LB, Rutter WJ (1980) Integration of hepatitis B virus sequences and their expression in a human hepatoma cell. Nature 286: 535–538
Evarts RP, Nagy P, Marsden E, Thorgeirson S (1987) A precursor-product relationship exists between oval cells and hepatocytes in rat liver. Carcinogenesis 8: 1737–1740
Farber E (1973) Carcinogenesis-cellular evaluation as a unifying thread: presidential address. Cancer Res 33: 2537–2550
Farber E (1980) Sequential analysis of liver cancer induction. Biochim Biophys Acta 605: 149–166
Farber E (1984) Pre-carcerous steps in carcinogenesis: their physiological adaptive nature. Biochim Biophys Acta 738: 171–180
Forrester K, Almoguera C, Han K, Grizzle WE, Perucho M (1987) Detection of high incidence of K-ras oncogenes during human colon tumorigenesis. Nature 327: 298–303

Fourel G, Trepo C, Bougueleret L et al. (1990) Frequent activation of N-*myc* genes by hepadnavirus insertion in woodchuck liver tumors. Nature 347: 294–298
Francke U, Riccardi VM (1979) Aniridia-Wilms' tumor association: evidence for specific deletion of 11p13 Cytogenet Cell Genet 23: 185–192
Friend SH, Dryja TP, Weinberg RA (1989) Oncogenes and tumor-suppressing genes. N Eng J Med 318: 618–622
Fu XX, Su CY, Lee Y, Hintz R, Biempica L, Snyder R, Rogler CE (1988) Insulin-like growth factor II expression and oval cell proliferation associated with hepatocarcinogenesis in woodchuck hepatitis virus carriers. J Virol 62: 3422–3430
Gilmour DS, Elgin SCR (1987) Localization of specific topoisomerase I interactions within the transcribed region of active heat shock genes by using the inhibitor comptothecin. Mol Cell Biol 17: 141–148
Grandgenett OP, Mumm SR (1990) Unraveling retrovirus integration. Cell 60: 3–4
Greene WC, Leonard J, Wang Y, Svetlik PB, Peffer NJ, Sodorski, JG, Rosen CA, Goh WC, Haseltine WA (1986) Transactivator gene of HTLV-I induces IL-2 receptor and IL-2 cellular gene expression. Science 232: 877–880
Gust ID, Burrell CJ, Coulepis AG et al. (1986) Taxonomic classification of human hepatitis B virus. Intervirology 25: 14–29
Hada H, Arima T, Togawa K, Okada Y, Morichika S, Nagashima H (1986) State of hepatitis B viral DNA in the liver of patients with hepatocellular carcinoma and chronic liver disease. Liver 6: 189–198
Halligan BD, Davis JL, Edwards KA, Liu LF (1982) Intra-and intermolecular strand transfer by HeLa DNA topoisomerase I. J Biol Chem 257: 3995–4000
Hayward WS, Neel BG, Astrin SM (1981) Acivation of a cellular oncogene by promoter insertion in ALV-induced lymphoid leukosis. Nature 290: 475–480
Heard JM, Fichelson S, Sola B, Martial MA, Varet B, Levy JP (1984) Multistep virus-induced leukemogenesis *in vitro*: description of a model specifying three stems within the myeloblastic malignant process. Mol Cell Biol 4: 216–220
Held WA, Mullins JJ, Kuha NJ, Gallagher JF, Gu GD, Gross KW (1989) T antigen expression and tumorigenesis in transgenic mice containing a mouse major urinary protein/SV40 T antigen hybrid gene. EMBO J 8: 183–191
Hino O, Shows TB, Rogler CE (1986) Hepatitis B virus integration site in hepatocellular carcinoma at chromosome 17:18 translocation. Proc Natl Acad Sci USA 83: 8338–8342
Hino O, Ohtake K, Rogler CE (1989) Features of two hepatitis B virus (HBV) DNA integrations suggest mechanisms of HBV integration. J Virol 63:2638–2643
Hixon DC, Allison JP (1985) Monoclonal antibodies recognizing oval cells induced in the liver of rats by N-2-fluorenylacetamide or ehtionine in a choline-deficient diet. Cancer Res 45: 3750–3760
Hixon DC, Faris RA, Thompson, NL (1990) An antigenic portrait of the liver during carcinogenesis. Pathobiology 10 (in press)
Hollinger B (1985a) Hepatitis B viruses. In: Fields BN et al. (eds) Virology. Raven, New York, pp 1396–1398
Hollinger B (1985b) Features of viral hepatitis: part I: clinical features. In (eds) Fields BN et al. Virology. Raven, New York pp 1418–1423
Hsu TY, Moroy T, Etiemble J, Louise A, Trepo C, Tiollais P, Buendia M-A (1988) Activation of c-myc by woodchuck hepatitis virus infection in hepatocellur carcinoma. Cell 55: 627–635
Huber BE, Thorgeirson SS (1987) Analysis of c-myc expression in human hepatoma cell line. Cancer Res 47: 3414–3420
Kehule AS, Laurer U, Meyer M, Caselmann WH, Hofschneider, Koshy R (1990) The preS2/S region of integrated hepatitis B virus DNA encodes a transcriptional transactivator. Nature 343: 457–461
Konopka AK (1988) Compilation of DNA strand exchange sites for nonhomologous recombination in somatic cells. Nucleic Acids Res 16: 1739–1758
Korba BE, Tennant BC, Gowans E et al. (1987) Infection and tissue tropism of WHV during the natural course of viral infection of woodchucks. In: Robinson W, Koike K, Will H (eds) Hepadnaviruses. Liss, New York, pp 419–428
Koshy R, Maupas P, Muller R, Hofschneider PH (1981) Detection of hepatitis B virus-specific DNA in the genomes of human hepatocellular carcinoma and liver cirrhosis tissues. J Gen Virol 57: 95–102
Koufos A, Hansen MF, Copeland NG, Jenkins NA, Lampkin BC, Cavenee WK (1985) Loss of heterozygosity in three embryonal tumours suggests a common pathogenetic mechanism. Nature 316: 330–334

Lee TH, Finegold MJ, Shen RF et al. (1990) Hepatitis B virus transactivator X protein is not tumorigenic in transgenic mice. J Virol 64: 5939–5947
Lombardi B (1982) On the nature, properties and significance of oval cells. In: Pani P, Feo F, Columbano A (eds) Cagliari. ESA, Rome, Italy, pp 36–56
Ludlow JJ, DeCaprio JA, Huang CM, Lee WW, Paucha E, Livingston DL (1989) SV40 large T antigen binds preferentially to an under phosphorylated member of the retinoblastoma susceptibility gene product family. Cell 56: 57–65
Marion PL, Oshiro LS, Regency DC, Scullard GH, Robinson WS (1980) A virus in Beechey ground squirrels that is related to hepatitis B virus of humans. Proc Natl Acad Sci USA 77: 2941–2945
Marion PL, Knight SS, Salazar FH, Popper H, Robinson WS (1983) Ground squirrel hepatitis virus infection. Hepatology 3: 519–527
Marion PL, Van Davelaar MJ, Knight SS, Salazar FH, Garcia G, Popper H, Robinson WS (1986). Hepatocellular carcinoma in ground squirrels persistently infected with ground squirrel hepatitis virus. Proc Natl Acad Sci USA 83: 4543–4546
Maruyama M, Shibuya H, Harada H, Hatakeyama M, Seiki M, Fujita T, Inoue J, Yoshida M, Taniguchi T (1987) Evidence for aberrant activation of the interleukin-2 autocrine loop by HTLV-1 encoded p40x and T3/T1 complex triggering. Cell 48: 343–350
Mason WS, Taylor JM (1989) Experimental systems for the study of hepadnavirus and hepatitis delta virus infections. Hepatology 9: 635–645
Mason WS, Seal S, Summers J (1980) Virus of Pekin ducks with structural and biological relatedness to human hepatitis B virus. J Virol 36: 829–836
McBurney MW, Reuhl KR, Ally AA, Nasipuri S, Bell JC, Craig J (1988) Differentiation and maturation of embryonyl carcinoma derived neurons in cell culture. J Neurosci 8: 1063–1073
McMahon J, Richard WL, del Campo AA, Song MK, Thorgeirson SS (1986) Differential effects of transforming growth factor B on proliferation of normal and malignant rat liver epithelial cells in culture. Cancer Res 46: 4665–4671
Mead JE, Fausto N (1989) Transforming growth factor may be a physiological regulator of liver regeneration of means of an autocrine mechanism. Proc Natl Acad Sci USA 86: 1558–1562
Milich DR, Chisari FV (1982) Genetic regulation of the immune response to hepatitis B surface antigen (HBsAg) 1 H-2 restriction of the murine humoral immune response to the a and d determinants of HBsAg. J Immunol 129: 320–420
Milich DR, McLachlan A (1986) The nucleocapsid of hepatitis B virus is both a T-cell-independent and a T-cell dependent antigen. Science 234: 1398–1401
Milich DR, Thornton GB, Neurath A (1985) Enhanced immunogenicity of the preS region of hepatitis B surface antigen. Science 228: 1195–1199
Milich DR, McLachlan A, Chisari FV, Kent SBH, Thornton GB (1986) Immune response to the preS (1) region of the hepatitis B surface antigen (HbsAg): a preS (1)-specific T cell response can bypass nonresponsiveness to the preS (2) and S regions of the HBsAg. J Immunol 137: 315–322
Milich DR, McLachlan A, Thornton GB, Hughes JL (1987) Antibody production to the nucleocapsid and envelope of the hepatitis B virus primed by a single synthetic T cell site. Nature 329: 547–549
Miller RH, Robinson WS (1986) Common evolutionary origin of hepatitis B virus and retroviruses. Proc Natl Acad Sci USA 83: 2531–2535
Minshull J, Blow JJ, Hunt T (1989) Translation of Cyclin mRNA is necessary for extracts of activated xenopus eggs to enter mitosis. Cell 59: 947–956
Mizusawa H, Taira M, Yaginuma K, Kobayashi M, Yoshida, E, Koike K (1985) Inversely repeating integrated hepatitis B virus DNA and cellular flanking sequences in the human hepatoma-derived cell line huSP. Proc Natl Acad Sci USA 82: 208–212
Morgan DO, Edman JC, Standring DN, Fried VA, Smith MC, Roth RA, Rutter WJ (1987) Insulin-like growth factor II receptor as a multifunctional binding proteins. Nature 329: 301–307
Moroy T, Marchio A, Etiemble J, Trepo C, Tiollais P, Buendia M-A (1986) Rearrangement and enhance depression of c-myc in hepatocellular carcinoma of hepatitis virus infected woodchucks. Nature 324: 276–279
Nagaya T, Nakamura T, Tokino T, Isurimoto T, Imai M, Mayumi T, Kamino K, Yamamura K, Matsubara K (1987) The mode of hepatitis B virus DNA integration in chromosomes of human hepatocellular carcinoma. Genes Dev 1: 773–782
Nakamura T, Teramoto H, Ichihara A (1986) Purification and characterization of a growth factor from rat platelets for mature parenchymal hepatocytes in primary cultures. Proc Natl Acad Sci USA 83: 6489–6493

Nakamura T, Tokino T, Nagaya T, Matsubara K (1988) Microdeletion associated with the integration process of hepatitis B virus DNA. Nucleic Acids Res 16: 4865–4873
Nakamura T, Nishizawa T, Hagiya M, Seki T, Shimoniski M, Sugimura A, Tashiro K, Shimizu S (1989). Molecular cloning and expression of human hepatocyte growth factor. Nature 342: 440–443
Ochiyda T, Fujiyama A, Fukushige S, Hatada I, Matsubara K (1986) Molecular cloning of an oncogene from a human hepatocellular carcinoma. Proc Natl Acad Sci USA 83: 4993–4997
Ogston W, Jonak GJ, Rogler CE, Astrin SM, Summers J (1982) Cloning and structural analysis of integrated woodchuck hepatitis virus sequences from hepatocellular carcinomas of woodchucks. Cell 29: 385–394
Omato M, Uchiumi K, Ito Y, Yokosuka O, Mori J, Terao K, Wei-Fa Y, O'Connell AP, London W, Okuda K (1983) Duck hepatitis and liver diseases. Gastroenterology 85: 260–267
Pitot HC, Sinca AE (1980) The stages of initiation and promotion in hepatocarcinogenesis. Biochim Biophys Acta 605: 191–203
Popper H, Roth L, Purcell RH, Tennant BC, Gerin JL (1987) Hepatocarcinogenicity of the woodchuck hepatitis virus. Proc Natl Acad Sci USA 84: 866–870
Radziwill G, Tucker W, Schaller H (1980) Mutational analysis of the hepatitis B virus P gene product: domain structure and RNase H activity. J Virol 64: 613–62
Raimondo G, Burk RD, Lieberman HM, Muschel J, Hadziyannis SJ, Will H, Kew MC, Dusheiko JM, Shafritz DA (1988) Interrupted replication of hepatitis B virus in liver tissue of HBsAg carriers with hepatocellular carcinoma. Virology 166: 103–112
Rall LB, Scott J, Bell G, Crawford RJ, Penschow JD, Niall HD, Coghlan J (1985) Mouse prepo-epidermal growth factor synthesis by the kidney and other tissues. Nature 313: 228–231
Rechler MM, Nissley PS (1985) The nature and regulation of the receptors for insulin-like growth factors. Annu Rev Physiol 47: 425–422
Reynolds SH, Stowers SJ, Maronpot PR, Anderson M, Aaronson SA (1986) Detection and identification of activated oncogenes in spontaneously benign and malignant HCC's of B6C3F1 mouse. Proc Natl Acad Sci USA 83: 33–37
Reynolds SH, Stowers SJ, Patterson RM, Maronpot PR, Aaronson SA, Anderson MW (1987) Activated oncogenes in B6C3F1 mouse liver tumors: implications of risk assessment. Science 237: 1309–1316
Rogler CE, Summers J (1982) Novel forms of woodchuck hepatitis virus DNA isolated from chronically infected woodchuck liver nuclei. J Virol 44: 852–863
Rogler CE, Summers J (1984) Cloning and structural analysis of integrated woodchuck hepatitis virus sequences from a chronically infected liver. J Virol 50: 832–837
Rogler CE, Sherman M, Su CY, Shafritz DA (1985) Deletion in chromosome 11p associated with a hepatitis B integration site in hepatocellular carcinoma. Science 230: 319–322
Romanus JA, Terrell JW, Yang YW, Nissley P, Rechler M (1986) Insulin-like growth factor carrier proteins in neonatal and adult rat serum and immunologically different. Endocrinology 118: 1743–1757
Ruley HE (1983) Adenovirus early version 1A enables viral and cellular transforming genes to transform primary cells in culture. Nature 304: 602–603
Sandgren EP, Quaife CJ, Pinkert CA, Palmiter RD, Brinster RL (1989) Oncogene-induced liver noeplasia in transgenic mice. Oncogene 4: 715–724
Scherer E (1989) Neoplastic progression during or in experimental hepato-carcinogenesis. Biochim Biophys Acta 738: 219–236
Sell S (1978) Distribution of alphafetoprotein and albumin containing cells in the livers of Fischer rats fed four cycles of N-2-fluorenylacetamide. Cancer Res 38: 3107–3113
Sell S, Salman J (1984) Light and electron microscopic autoradiographic analysis of proliferating cells during the early stages of chemical carcinogenesis in the rat induced by feeding N-2-fluorenylacetamide in a choline deficient diet. Am J Pathol 114: 287–300
Sell S, Osborn K, Leffert HL (1981) Autoradiography of "oval cells' appearing rapidly in the livers of rats fed N-2 fluorenylacetamide in a choline devoid diet. Carcinogenesis 2: 7–14
Sell S, Hunt JM, Knoll BJ, Dunsford HA (1987) Cellular events during hepatocarcinogenesis and the question of premalignancy. Adv Cancer Res 48: 37–111
Seto E, Mitchell PJ, Yen TSB (1990) Transactivation by the hepatitis B virus X protein depends on AP-2 and other transcription factors. Nature 344: 72–74
Shafritz DA, Shouval D, Sherman HI, Hadziyannis SJ, Kew MC (1981) Integration of hepatitis virus DNA into the genome of iver cell in chronic liver disease and hepatocellular carcinoma. N Engl J Med 305: 1067–1073

Shaul Y, Ziemer M, Garcia PD, Crawford R, Hsu H, Valenzuela P, Rutter WJ (1984) Cloning and analysis of integrated hepatitis virus sequences from a human hepatoma cell line. J Virol 51: 776–787

Shih C, Burke K, Chow MJ, Zeldis JB, Yang CS, Lee CS, Isselbacher KJ, Wands JR, Goodman HM (1987) Tight clustering of human hepatitis B virus integration sites in hepatomas near a triple-stranded region. J Virol 61: 3491–3498

Snyder RL, Summers J (1980) Woodchuck hepatitis virus and hepatocellular carcinoma from viruses in naturally occurring cancers. In: Essex M, Todaro G, Zur Hausen H (eds), Cold Spring Harbor Conference of Cell Proliferation, vol 7. Cold Spring Harbor Laboratory, Cold Spring Harbor, pp 447–457

Soares MB, Ishii DN, Efstratiadis A (1985) Developmental and tissue-specific expression of a family of transcripts related to rat insulin-like growth factor II mRNA. Nucleic Acids Res 14: 1119–1134

Soares MB, Turken A, Ishii D, Mills L, Epeskopou V, Cotter S, Zeitlin S, Efstratiadis A (1986) Rat insulin-like growth factor II gene, a single gene with two promoters expressing a multitranscript family. J Mol Biol 192: 737–753

Sobczak J, Mechti N, Francoise-Tourmer M, Blanchard JM, Duguet M (1989) C-myc and C-fos gene regulation during mouse liver regeneration. Oncogene 4: 1503–1508

Spandau DF, Lee CH (1988) Trans-activation of viral enhances by the hepatitis B virus X protein. J Virol 62: 427–434

Sprengel R, Kaleta EF, Will H (1988) Isolation and characterization of a hepatitis B virus endemic in herons. J Virol 62: 3832–3839

Summers J, Smolec J, Snyder R (1978) A virus similar to human hepatitis B virus associated with hepatitis and hepatoma in woodchucks. Proc Natl Acad Sci USA 75: 4533–4537

Summers J, Smolec JM, Weiner BG, Kelly TJ, Tyler GW, Snyder RL (1980) Hepatitis B virus and woodchuck hepatitis virus are members of a novel class of DNA viruses from viruses. In: Essex M, Todaro G, Zur Hansen H (eds) Naturally occurring cancers book A, pp 459–469 (Cold Spring Harbor Conference in Cell Proliferation, vol 7)

Takada S, Koike K (1989) Activated N-ras gene was found inhuman hepatoma tissue but only in a small fraction of the tumor cells. Oncogene 4: 189–193

Takada S, Goth Y, Hayasti S, Yostida M, Koike K (1990) Structural rearrangement of integrated hepatitis B virus DNA as well as cellular flanking DNA is present in chronically infected hepatic tissues. J Virol 64: 822–828

Thompson NL, Mead JE, Braun L, Goyette M, Shank PR, Fausto N (1986) Sequential protooncogene expression during rat liver regeneration. Cancer Res 46: 3111–3117

Tokino T, Fukushige S, Nakamua T, Nagaya T, Murotsu T, Shiga K, Aoki N, Matsubara K (1987) Chromosomal translocation and inverted duplication associated with integrated hepatitis B virus in hepatocellular carcinoma. J Virol 61: 3848–3854

Twu JS, Robinson WS (1989) Hepatitis B virus X gene can transactivate heterologous viral sequences. Proc Natl Acad Sci USA 86: 2046–2050

Twu JS, Schloemer RH (1987) Transcriptional transacting function of hepatitis B virus. J Virol 61: 3448–3453

Ueno T, Takahashi K, Matsuguchi T, Endo H, Yamamoto M (1988a) Transcriptional deviation of the rat insulin-ike growth factor II gene initiated at three alternative leader—exons between neonatal tissues and acites hepatomas. Biochim Biophys Acta 950: 411–419

Ueno T, Takahashi K, Matsuguchi T, Ikejiri K, Endo H, Yamamoto M (1988b) Reactivation of rat insulin-like growth factor II gene during hepatocarcinogenesis. Carcinogenesis 9: 1779–1783

Wang HP, Rogler CE (1988) Deletions in chromosome arms 11p and 13q in primary hepatocellular carcinomas. Cytogenet Cell Genet 48: 72–78

Wang HP, Rogler CE (1991) Topoisomerase I mediated hepadnavirus DNA integration in vitro. J Viral (in press 1991)

Wang J, Chenwesse X, Henglein B, Brechot C (1990) Hepatitis B virus integration in a cyclin A gene in a hepatocellular carcinoma. Nature 343: 555–557

Wollersheim M, Debelka U, Hofschneider PH (1988) A transactivating function encoded in the hepatitis B virus X gene is conserved in the integrated state. Oncogene 3: 545–554

Wong S, Winchell LF, McCune BK, Earp H, Teixido S, Massague J, Herman B, Lee DC (1989) The TGF-precursor expressed on the cell surface binds to the EGF receptor on adjacent cells leading to signal transduction. Cell 56: 495–506

Wong-Staal F, Gallo RC (1985) Human T-lymphotropic retroviruses. Nature 317: 395–403

Wu J, Zhou Z-Y, Judd A et al. (1990) The hepatitis B virus-encoded transcriptional activator Hbx appears to be a novel protein serine/threonine kinase. Cell 63: 687–695

Yaginuma K, Kobayashi M, Yoshida E, Koike K (1985) Hepatitis B virus integration in hepatocellular carcinoma DNA: Duplication of cellular flanking sequences at the integration site. Proc Natl Acad Sci USA 82: 4458–4462

Yaginuma K, Kobayashi H, Kobayashi M, Morishima T, Matsuyama K, Koike K (1987) Multiple integration site of hepatitis B virus DNA in hepatocellular carcinoma and chronic active hepatitis tissues from children. J Virol 61: 1808–1813

Yang DY and Rogler CE (1991) Insulin-like growth factor (IGF-II) and mannose-6-phosphate/IGF-II receptor expression in neoplastic nodules and hepatocellular carcinomas: A potential autocrine role in malignant progression in the liver. Cancer Research (in press)

Yokosuka O, Omata M, Zhou YZ, Imazeki F, Okuda T (1985) Duck hepatitis B virus DNA in liver and serum of chinese ducks: integration of viral DNA in a hepatocellular carcinoma. Proc Natl Acad Sci USA 82: 5180–5184

Zahm P, Hofschneider PH, Koshy R (1988) The HBV X-ORF encodes a transactivator: a potential factor in viral hepatocarcinogenesis. Oncogene 3: 169–177

Human Hepatitis Delta Virus

J. M. TAYLOR

Institute for Cancer Research, Fox Chase Cancer Center, 7701 Burholme Avenue, Philadelphia, PA 19111, USA

Current Topics in Microbiology and Immunology, Vol. 168

1 Introduction

1.1 Objectives

In the past few years significant progress has been made in understanding the structure and replication of hepatitis delta (HDV). Some of this work has been recently reviewed (BONINO et al. 1990; HOOFNAGLE 1989; TAYLOR 1990a, b). The present chapter will emphasize those areas in which our perspective has recently changed and those questions about HDV which are (almost) accessible to laboratory research.

1.2 Discovery

RIZZETTO et al. (1977) obtained the first clue to the existence of HDV while studying liver biopsies from patients known to be infected with hepatitis B virus (HBV) but experiencing an unexpectedly high level of liver diease activity. They demonstrated by immunofluorescence microscopy, using antibodies obtained from these patients, the presence of a novel nuclear antigen that was not found in other individuals infected with HBV. This antigen, designated delta, was subsequently proven by experimental transmission studies to be part of an infectious agent, frequently referred to as HDV, that had superinfected this HBV-infected population. RIZZETTO et al. (1980) first demonstrated transmission of this agent to chimpanzees, and PONZETTO et al. (1984) subsequently transmitted this virus to woodchucks.

In both natural and experimental infections it has been observed that HDV replication is dependent upon simultaneous infection by an hepadnavirus, such as HBV in primates or woodchuck hepatitis virus (WHV) in woodchucks. As will be discussed below, the hepadnavirus is necessary to provide an envelope for the genome of HDV. Therefore, according to the definitions of DIENER and PRUSINER (1985) HDV is not a virus. It is actually "subviral" and should probably be called a "satellite RNA" of the hepadnavirus; however, this designation has not yet been widely accepted.

1.3 Natural Infections

Transmission of HDV is apparently not different from that of its helper hepadnavirus, that is, via contaminated blood or blood products. Such infections can be divided into two main types: "coinfections," in which the HDV and hepadnavirus are introduced simultaneously, and "superinfections," in which the HDV is introduced after a chronic hepadnavirus infection has become established. Both types of infection have a significant chance of producing a fatal hepatitis; it would seem that superinfections can be more dangerous, with up to a 20% chance of fulminant hepatitis if the patient already has a chronic active hepatitis as the time of superinfection. Survivors of coinfections frequently clear the HDV as well as the HBV infection. In contrast, those surviving a superinfection almost invariably go on to develop a chronic HDV infection superimposed upon the chronic HBV infection (HOOFNAGLE 1989; NEGRO et al. 1988). There is evidence that chronic HDV infections are associated with increased liver damage, but there is no evidence of an increased risk of developing hepatocellular carcinoma over that due to chronic HBV infection alone.

With the hepadnaviruses there is a liver pathogenicity that seems to be largely mediated by the immune response of the host. However, it has been generally asserted that this is not the case for the pathogenic effects of HDV infections. This is consistent with the claims by POPPER et al. (1983) that the liver of an HDV-infected individual reveals characteristic cytopathic effects. It must be noted that there are some recent claims for an immune involvement in the pathogenicity of HDV infections (referred to in BONINO et al. 1990). The ability of HDV replication to interfere with HBV replication is discussed in Sect. 7 below.

1.4 Prevention and Treatment

It has been shown that immunization against the surface antigen of the helper virus, HBV, can protect against infection with HDV. This is not surprising since, as explained in Sect. 2, the HBV surface antigen is used as part of the envelope of HDV. Experiments are underway to determine whether prior immunization using recombinant delta antigen confers any protection.

Unfortunately, once an HDV infection has been established there is as yet no efficacious antiviral treatment procedure. The replication of HDV, as reflected by the presence in the blood stream of liver enzymes, indicative of ongoing liver damage, and by circulating virus, can be temporarily suppressed by treatments with interferon, but following withdrawal from treatment and sometimes even earlier, HDV replication and associated liver damage resume (HOOFNAGLE 1989). This result is very similar to what has been found following interferon treatment of individuals with HBV and hepatitis C virus (HCV) infections (DAVIS et al. 1989).

1.5 Epidemiology

With the availability of diagnostic tests for the delta antigen and for antibodies directed against this antigen, it has become clear that HDV infections are widespread. HOOFNAGLE (1989) has pointed out that HDV occurs in three epidemiological patterns. (1) An endemic pattern is found in certain populations (e.g., the Mediterranean area and the Middle East) with a high level of chronic HBV carriers. (2) Localized epidemics have occurred amongst isolated communities (e.g., northern South America and central Africa) with a high level of chronic carriers. (3) Minor epidemics have occurred amongst certain high-risk groups (e.g. intravenous drug users, patients who receive multiple transfusions or blood products). HDV, like its natural helper virus HBV and also like human immunodeficiency virus (HIV), is spread via contaminated blood. It has become clear from re-examination of stored liver samples that HDV outbreaks were present (e.g. in South America) at least as early as 1937 (BUITRAGO et al. 1986).

There has been suggestive evidence for strain variations between different HDV isolates, i.e., there are examples of what might be interpreted as more pathogenic and of milder to relatively non-pathogenic HDV outbreaks. What does all this mean? The variations in pathogenicity may occur for different reasons, not all of which are attributable to variations in the HDV itself. For example, it is not yet clear whether more severe outbreaks of HDV were superimposed on a clinically more significant level of HBV-associated liver disease.

2 Virions

2.1 Structure

In natural HDV infections both the HDV and the helper hepadnavirus have been shown to replicate in hepatocytes. Both types of virus particles are released into the blood of the infected host. As reviewed in this volume (chapter by D. GANEM) and elsewhere (GANEM and VARMUS 1987; SCHODEL et. al. 1989), the hepadnavirus particles include the infectious particles, the so-called Dane particle, which is about 42 nm in diameter and contains a 27-nm core structure. There is also an excess of two types of noninfectious hepadnavirus particles: These are the so-called surface antigen particles, present as both 22-nm spheres and as filaments that are 22 nm in diameter and up to several hundred nanometers in length. The HDV particles are present along with the various hepadnavirus particles, and all have an outer envelope that is substituted with the surface antigen proteins encoded by the hepadnavirus.

The HDV particles can be distinguished from the hepadnavirus particles, by differences in density and sedimentation values (BONINO et al. 1984, 1986). Initial isolation studies made use of the fact that the delta antigen was within the HDV virion and could be readily assayed following release by treatment with a nonionic detergent. Later studies made use of assays for the HDV RNA genome, also present within these particles (see Sect. 2.2). Other studies have made use of immuno-electron microscopy (RIZZETTO et al. 1980). A recent study using filtration by HE et al. (1989) indicates that the HDV particles are heterogeneous in diameter, with a mean of 38 nm. The studies of BONINO et al. (1984) indicated that the particles lacked a demonstrable core structure or even a ribonucleoprotein structure. They observed that the HDV RNA and antigen, when released by mild treatment with nonionic detergent, showed no consistent association. This could have been because of an artifact of preparation and/or disruption, since, as discussed later in Sect. 4, it is now known that the HDV RNA and delta antigen can make a stable interaction.

In the serum of an infected host the titer of HDV can be as high as 2×10^{11} infectious units per milliliter (PURCELL et al. 1987). The actual number of particles containing HDV RNA can be similar, suggesting that the specific infectivity, under ideal conditions, may be close to one.

2.2 Genome and Antigenome

Inside the HDV particle is a small single-stranded RNA molecule, about 1700 bases in length, which is the genome. It was shown independently by WANG et al. (1986), KOS et al. (1986), and CHEN et al. (1986) that this RNA is unusual in that it is circular in conformation. Also, WANG et al. (1986) showed both by electron microscopy and nucleotide sequencing that the RNA has the ability to fold onto itself by base pairing to form an unbranched rod structure. KUO et al. (1988a) calculated that the negative free energy of this folding was as high as 805 kCal/mol. Such structural features, as discussed in Sect. 3.3, are reminiscent

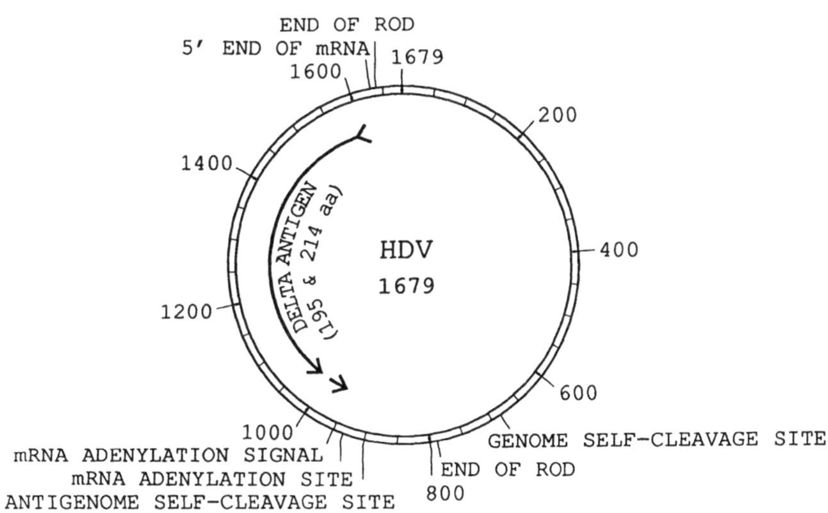

Fig. 1. Sequence features on the genome and antigenome of HDV. The 1679-b sequence of KUO et al. (1988a) is shown with the genome oriented clockwise, and the antigenome superimposed and shown anticlockwise. The predicted ends of the unbranched rod structure are at 793 and 1636. The sites of self-cleavage on the genome and antigenome are at 685/686 and 900/901, respectively. On the antigenome is indicated the coding regions for the small (1598–1010, 195 aa) and large (1598–953, 214 aa) forms of the delta antigen, along with the 5′ end (1631) and polyadenylation site (1934) of the mRNA. The polyadenylation signal, AAUAAA, shown to be essential for the expression of the delta antigen is located at 954–949.

of what others have demonstrated for the genomes of certain of the subviral agents of plants (SÄNGER 1987). From both hybridization studies and nucleotide comparisons it would seem that there is no significant sequence relatedness between the hepadnaviruses and HDV (WANG et al. 1987). Also, there is no proven relatedness between the HDV and a host genome.

Inside infected cells it is possible to detect not only the HDV genome but also relatively lower amounts of a complementary RNA, which is called the "antigenome." From the nucleotide sequences of both the genome and antigenome it is possible to predict several potential open reading frames for HDV proteins (WANG et al. 1986, 1987; MAKINO et al. 1987; KUO et al. 1988a). However, at this time, the only one that is known to function during infection is that encoding the delta antigen (WEINER et al. 1988; CHANG et al. 1988). This reading frame, located on the antigenomic RNA (see Fig. 1), is expressed via translation of a specific mRNA that is considerably shorter than the antigenome (see Sect. 4.3 and Fig. 2). In early studies of HDV, only genomic RNA was detected within virions, and antigenomic RNA was estimated to be at least 100-fold less abundant (CHEN et al. 1986). However, by using the sensitivity of the polymerase chain reaction, it has become possible recently to detect antigenomic RNA in purified virions (G. LUC and J. TAYLOR, unpublished observations).

3 Genome Replication

At the outset it should be made clear that whereas hepadnaviruses replicate via reverse transcription of an RNA intermediate (SUMMERS and MASON 1982), HDV does not (CHEN et al. 1986). As will be explained, HDV replicates via RNA-directed RNA synthesis. We will begin with a summary of the experimental evidence and then consider specific models that may explain the replication strategy of this HDV genome.

3.1 Experimental Systems

3.1.1 Infected Animals

As mentioned in Sect. 1.3, HDV infection occurs naturally in humans, and experimental transmission has been achieved in chimpanzees and woodchucks. In all such studies, HDV has been found in association with a hepadnavirus, be it HBV for primates or WHV for woodchucks. One might thus expect that HDV would also replicate in ducklings infected with duck hepatitis B virus (DHBV). However, it is puzzling that contrary to such an expectation and in spite of preliminary results (PONZETTO et al. 1987), several laboratories have been unsuccessful in achieving this goal.

With the three available host systems (humans, chimpanzees, and woodchucks), it has been possible to study not only the replication of HDV but also the interactions between HDV and the helper hepadnavirus, as discussed in Sect. 3.5. In the experimental superinfections of chimpanzees and woodchucks it has been clearly shown that the peak of HDV viremia occurs about 5–7 weeks postinoculation with HDV (PONZETTO et al. 1984). This peak arises and promptly disappears, all within several days. This can be a major experimental problem unless one is able to take multiple serum and even biopsy samples, in order to catch the infection at its peak.

3.1.2 Infected Hepatocytes

TAYLOR et al. (1987b) and later CHOI et al. (1988) have demonstrated that primary cultures of woodchuck hepatocytes are susceptible to infection by HDV that had been grown in a woodchuck that was chronically infected with WHV, and SUREAU et al. (1991) have shown that primary cultures of chimpanzee hepatocytes can also support HDV replication.

Experiments with woodchuck hepatocytes reveal that the cultured cells do not need a prior infection with WHV in order to synthesize HDV RNA replicative intermediates, suggesting that WHV does not contribute to this process (TAYLOR et al. 1987b). This result supports the hypothesis that the only role of a hepadnavirus in HDV replication is to provide the envelope of the mature virion. Curiously, in vivo, coinfection by HDV causes significant suppression of hepadnavirus replication (see Sect. 7).

The results of in vitro infection of primary hepatocyte cultures have so far confirmed the host specificity reported for infected animals. They have also been valuable for in situ hybridization to localize sites of HDV RNA accumulation and for studying other aspects of the time course of HDV infections (TAYLOR et al. 1987b). Using the same system PETCU et al. (1988) and later CHOI et al. (1989) have shown that suramin acts on both HDV and HBV and blocks virus entry. CHOI et al. (1989) have also studied the effects of α-amanatin, acyclovir, and ribavarin. However, neither animal nor tissue culture studies have so far clarified one of the major questions concerning HDV replication, namely, the nature of the polymerase involved in HDV RNA synthesis (see Sects. 3.1.6, 3.8, and 8). There is the possibility that some of the effects of acyclovir and ribavarin as reported by CHOI et al. may have been indirect, acting on the survival of the hepatocyte cultures rather than on HDV replication.

3.1.3 Cells Transfected with cDNA

It has been possible to assemble a full-length cDNA clone of the HDV genome and to insert multimers of this clone into a eukaryotic expression vector. KUO et al. (1989) inserted a trimer of HDV into pSVL so that genomic HDV RNA is expressed via DNA-dependent transcription directed by the late promotor of SV40 to yield a trimer of genomic HDV RNA. Transfection of cell lines with this clone leads to subsequent synthesis of the monomeric HDV genome via RNA-dependent RNA synthesis. Such studies provided the final proof that there is no need for HBV sequences or functions in this process, thereby confining the roles of HBV in complete cycles of HDV replication to packaging and virus entry. The transfection studies also demonstrated HDV RNA-dependent RNA synthesis in both liver and nonliver cell lines, including mouse fibroblasts, thus showing that there is no need for liver-specific functions (CHAO et al. 1991). Independent studies by MACNAUGHTON et al. (1990) have demonstrated that an HDV cDNA clone can also be constructed so as to initiate genome replication starting with the DNA-directed transcription of antigenomic rather than genomic HDV RNA. In addition, these researchers have established cell lines stably transfected with an HDV cDNA clone.

KUO et al. (1989) have used the transfection approach to establish that the presence of the delta antigen is essential for the replication of the HDV genome. More recently, this approach has been applied to study spontaneous mutation of the HDV genome and the mechanism of synthesis of the mRNA for the delta antigen. These latter studies are described in Sects. 4.3 and 4.4, respectively.

It is expected that future applications of transfection technology will continue to be informative. For example, this should be an ideal system with which to determine the HBV-specific and HDV-specific requirements for assembly and release of HDV virions. It should also make it possible to clarify the mechanism by which HDV replication interferes with that of HBV (see Sect. 7).

3.1.4 Animals Transfected with cDNA

After showing that a cDNA clone of HDV could initiate genome replication in transfected cells, it was important to know whether such DNA could initiate genome replication in an experimental animal. SUREAU et al. (1989) used the cDNA construct of KUO et al. (1989) described above and injected the DNA directly into the liver of a chimpanzee that was already chronically infected with HBV. This transmission experiment was successful. After 6 weeks the chimpanzee developed a full-blown HDV infection with all the expected serum markers indicative of liver damage and genome replication. There was also a clear transient suppression of HBV replication, as indicated both by a drop in the amounts of HBV in serum and by a decrease in HBV replicative intermediates in liver biopsy material. This study, unlike those described in the two previous sections, showed a completed cycle of HDV replication, in that progeny HDV was released. Surprisingly, the nucleotide sequence of the HDV produced in the chimpanzee had diverged considerably from that of the injected cDNA clone. The mechanism and significance of these changes is discussed in Sect. 4.4.

3.1.5 Cell Transfected with RNA

Recently GLENN et al. (1990) have observed HDV RNA synthesis following the introduction of in vitro synthesized HDV RNA into tissue culture cells. They first made use of a special delivery system previously used to deliver specific proteins and mRNAs to cultured cells. Briefly, a line of mouse fibroblasts that had been transfected with a cDNA clone of the influenza virus hemagglutinin and expressed the influenza protein at the cell surface was used as a recipient of liposomes containing HDV RNA. To achieve efficient uptake of the liposomes, GLENN et al. (1990) took advantage of the procedure described by ELLENS et al. (1989), in which liposomes substituted with glycophorin act as a ligand for the hemagglutinin expressed on the surface of the recipient cells. The liposomes were assembled to contain many copies of HDV genomic RNA. This RNA had been first transcribed in vitro with phage T7 RNA polymerase; it contained exactly one copy of HDV sequences but was flanked by short sequences from the transcription vector. After adsorption of the liposomes to the cultured cells, there was a brief treatment at a lowered pH so as to achieve fusion and release of the RNA into the recipient cells. This first experiment failed; however, successful induction of HDV RNA synthesis was obtained by slightly changing the recipient cells. Specifically, the recipient cells were stably transfected with a eucaryotic vector expressing the delta antigen, prior to introduction of HDV RNA. This study is also consistent with the cDNA transfections of KUO et al. (1989) and demonstrates that the delta antigen is required at one or more crucial steps in HDV genome replication. The finding of GLENN et al. in addition leads to the hypothesis that the delta antigen present within the virus particles may be needed for the initiation of HDV genome replication.

One important question, raised by the studies of GLENN et al. (1990), was how the linear unit length HDV RNA placed in the liposomes was effectively "healed" so as to allow what seems to be an otherwise normal genome replication. Cloning and sequencing of the healed progeny RNA showed that at least the majority of the progeny molecules were repaired so as to remove precisely the few nondelta sequences located at the ends of the infecting molecule. The enzymology of this healing reaction has not yet been explained.

3.1.6 Future Systems

What is really needed at this time is an in vitro system that supports the complete replication of the HDV genome. However, as discussed below (Sect. 3.8), even with a nuclear extract from infected cells, it should be possible to use nuclear run-on experiments to determine the polymerase(s) involved in HDV RNA-directed RNA synthesis. It may also be possible to use extracts of cells expressing only the delta antigen and then add HDV RNAs and observe replication in infected livers. Many informative experiments can be envisaged.

3.2 Observed RNA Species

In the above we have described five different systems that have been used to study HDV genome replication. At least as far as the characterization of RNA species is concerned, there are no significant differences between these systems.

The consensus picture is as follows:

1. There is no detectable HDV-related DNA present during HDV genome replication in infected livers.
2. The majority of genomic and antigenomic RNAs are usually unit length, i.e. about 1.7 kb.
3. The most abundant unit length RNA species is genomic, with approximately 5–20% as much antigenomic RNA. The amount of genomic RNA can be as much as 300 000 copies per average liver cell.
4. About half of the unit length RNAs are circular. The remainder are linear but do not appear to be opened at any specific location(s) on the genome (Kuo et al. 1988b), suggesting that linearization may be an artifact of the RNA extraction procedures.
5. There are minor amounts of other species that are greater than unit length. Some of these appear to be approximately dimers and to a lesser extent, trimers. As with the linears, about half of these species behave electrophoretically as if they were circular.
6. Additional minor species are present which are less than unit length. Such species have not yet been proven to be linear and/or circular. In some infections, especially of woodchucks, they can be more abundant than the unit length species (see Sect. 6).
7. All of the above-mentioned species seem to be located primarily in the nucleus of infected cells. This is supported by cell fractionation studies and also by in situ hybridization (Taylor et al. 1987b; Gowans et al. 1988). However, there is additional evidence for a species largely localized to the cytoplasm. This is the putative mRNA for the delta antigen, a poly(A)-containing, 800-b, antigenomic species (Hsieh et al. 1990; Jilbert et al. 1991; and Sect. 4.3).
8. A final species of HDV-related RNA is a possible replicative structure. Chen et al. (1986) and Taylor et al. (1987b) have presented evidence for the isolation of HDV RNA structures with significant amounts of intermolecular double-strandedness. As Weissmann (1989) points out, such structures are usually artifacts of extraction when isolated from cells replicating RNA viruses. However, it is worthwhile noting that Gowans et al. (1988) have also obtained evidence of double-stranded HDV RNAs by in situ hybridization. They found that little if any antigenomic RNA could be detected in the nuclei of infected cells unless the fixed cells were additionally submitted to a procedure which would denature double-stranded RNA structure; in contrast, genomic RNA, although capable of forming highly base-paired rod structures, was readily detected.

This summary of the HDV-related RNAs seen in infected cells is very much a partial story, because it does not include the involvement of the delta antigen. What then does the delta antigen do to facilitate HDV genome replication? As discussed in Sect. 4 it may have several functions. It has been known from the initial work of Rizzetto et al. (1977) that in the infected cell the majority of delta antigen is localized to the nucleus. Subsequent studies have complicated this matter. In certain transfection studies of CHANG et al. (1988) and KUO et al. (1989) the antigen was specifically located in the nucleolus, and yet in control studies with infected cells the antigen was more generally distributed over the whole nucleus. The significance of this difference remains unclear. Another complication, discussed in Sects. 4.4 and 6, is the modification of the coding region, so as to allow the synthesis of a second form of the antigen. Before attempting to resolve these data into a model of genome replication it will be helpful to see, as discussed in the following section, what can be learned from analogies between HDV and other systems.

3.3 Analogies to HDV in Other Systems

3.3.1 Pathogenic Agents of Plants

As reviewed elsewhere (TAYLOR et al. 1989; SYMONS 1989b), several features of HDV initially indicated that these might be an analogy between HDV and certain subviral pathogenic RNA agents of plants. These were: (1) HDV is subviral and needs a hepadnavirs helper for progeny virus production; (2) the RNA genome is not only single-stranded but also very small, actually smaller than that of any other known animal virus; (3) the HDV genome is circular in conformation; (4) the RNA has the ability to fold on itself by intramolecular base-pairing to form an unbranched rod structure; and (5) some of the HDV-related RNA species detected in infected tissues have structures consistent with RNA-directed RNA synthesis via a rolling-circle model (see Sect. 3.4). These five properties are shared with certain of the subviral plant agents, but the analogy is not strict. Nevertheless, the analogy between delta and these plant pathogens has predictive value. The two best examples of this were the demonstrations that HDV RNA, like that of the plant agents, can undergo in vitro self-cleavage and self-ligation (see Sects. 3.6 and 3.7, respectively).

In addition to the above, there is another possibly important similarity between HDV and the plant agents. As mentioned earlier, we know by cell fractionation and in situ hybridization that the HDV genome is replicated in the nucleus. In addition, by using immunofluorescence microscopy to detect the delta antigen, it has been shown that there are occasions when the location of delta antigen within the nucleus seems to be to the nucleus rather than the nucleoplasm. These results can be compared with the studies reported by HARDERS et al. (1989) which indicate that one of the plant viroids replicates in the nucleolus.

Despite the similarities, it is also true that the replication of HDV and the plant agents differ significantly in many ways. The major example is the requirement of the delta antigen for HDV RNA synthesis. Other details include a special mechanism of mRNA synthesis and regulation (as discussed in Sects. 4.3 and 5).

3.3.2 Introns

It has been pointed out previously by DINTER-GOTTLIEB (1986) and HADIDI (1986) that there may be some evolutionary relationship between certain of the plant agents and group I introns. More recently DIENER (1989) has even speculated the reverse, that the introns may have evolved from agents like viroids. Do these analogies have any relevance in our attempts to understand HDV? At this point the answer is not clear, but it is interesting, as explained in Sect. 3.6, that the self-cleavage of antigenomic HDV RNA shows some similarities to the removal of introns from mRNA precursors.

3.3.3 *7S* L RNA

Recently there has been a burst of interest as to the possible relationship between the 7S RNA component of the signal recognition particle and certain of the viroids and even HDV. HAAS et al. (1989) first pointed out a patch of sequence homology between certain viroids and *7S* L RNA. They and later SYMONS (1989a) noted that the homology was in the region of the viroid sequence that is considered to be a determinant of pathogenicity. NEGRO et al. (1989) and YOUNG and HICKE (1990) have claimed a homology between an important region on the HDV genome and the *7S* L RNA. The region of homology partially overlaps the site on the HDV genome at which in vitro self-cleavage has been shown to occur (as discussed in Sect. 3.6). This has led to speculations regarding the mechanism of pathogenicity of HDV, for example, that HDV might be able to reduce the amount of functional signal recognition particles via cleavage of *7S* L RNA. At present this proposal is highly speculative. Moreover, doubts can be raised about the significance of the original sequence comparisons (TAYLOR et al. 1990b). The answers will have to come from direct experimental evaluation.

3.4 Rolling-Circle Model

The concept of a rolling-circle model was introduced to explain the replication of viruses that were thought to have a circular intermediate in the synthesis of progeny nucleic acid (BROWN and MARTIN 1965). BRANCH and ROBERTSON (1984) applied such models to explain the replication of certain of the pathogenic plant RNAs; they even invoked a single- and double-circle model. SÄNGER (1987) has reviewed this and three related models for viroid replication. We have invoked another adaptation to explain the replication of HDV (TAYLOR et al. 1989). Recently (TAYLOR 1990b), a more cautious approach to this model has been advocated, largely for the reason that the model contains many highly speculative areas.

3.5 Role of the Hepadnavirus

From the cDNA transfection studies of Kuo et al. (1989) and the liposome studies of GLENN et al. (1990), as discussed in Sect. 3.1.3, it is clear that no HBV function or sequence is needed to obtain HDV genome reproduction. However, it is necessary to keep clear the distinction between the full cycle of HDV replication and the more limited concept of genome replication. We are tempted to conclude that the functions of the hepadnavirus are not needed for genome replication, and that the hepadnavirus is only needed for the assembly of new virions and later for reinfection into new host cells. This conclusion may need an additional qualification: It remains possible that in the liver of an infected animal even HDV RNA synthesis may depend upon one or more functions provided a hepadnavirus. A precedent for this distinction between tissue culture and animal studies is seen, for example, in the case of the thymidine kinase gene of herpes simpex virus, which is apparently not needed for virus replication in cultured cells but which in vivo seems to function as a facilitator of virus replication (FIELD and WILDY 1978).

3.6 Self-Cleavage

one of the major predictions of the analogy between HDV and the plant agents was that HDV RNA might demonstrate an in vitro ability to self-cleave. This was shown by SHARMEEN et al. (1988) and subsequently confirmed by WU et al. (1989) and by BRANCH et al. (1991). There is one site of self-cleavage on the HDV genomic RNA and one on the antigenomic RNA (see Fig. 1). These two sites are very similar in sequence, consistent with the extensive intramolecular pairing of the HDV genome. The cleavage reaction is dependent upon a divalent metal ion, such as magnesium. The reaction is also dependent upon temperature. The reaction is a transesterification, producing a 5′-OH and a cyclic 2′,3-monophosphate. KUO et al. (1988b) have elucidated the minimum amounts of contiguous sequence essential for self-cleavage; only about 100 bases are needed, with the cleavage site located near to the 5′-terminus.

For HDV, the self-cleavage of the antigenomic strand is particularly intriguing. The reason is because of the close proximity of several important *cis*-acting sequences, as indicated on Fig. 1. Consider the following: 56 bases 3′ of the termination codon for the delta antigen (see Sect. 4.3) is a polyadenylation signal, AAUAAA. This site is essential for the in vivo expression of the mRNA for the delta antigen (HSIEH et al. 1990). Some 20 bases further 3′ is the site at which poly(A) is added to make this mRNA. Then, 34 bases further 3′ is the self-cleavage site, which KUO et al. (1988b) have shown depends for its in vitro activity on 3′ and 5′ flanking sequences. Not only is there this close-packing, but the 16 bases 5′ of the self-cleavage site are all pyrimidines, analogous to the intron-exon border of mRNA introns. And, for the total of 34 bases between the adenylation site and the self-cleavage site there is only one adenosine, making the adenylation site somewhat like the branch point of a mRNA intron (SMITH et al. 1989).

With the self-cleaving RNAs of plants it has been possible not only to define the domain necessary for self-cleavage (BUZAYUN et al. 1986; FORSTER and SYMONS 1987; PRODY et al. 1986) but also to go further and separate out a catalytic domain from a substrate domain (UHLENBECK 1987). Such studies of HDV RNA are being carried out by several laboratories. To date, the RNA structure that induces self-cleavage of HDV has not been defined (LAI et al. 1991; BELINSKY and DINTER-GOTTLIEB 1991). As reviewed by SYMONS (1989b), amongst those RNAs of plant agents that do undergo self-cleavage, two types of structures have been defined: a so-called "hammer-head," which is found in many of these agents, and a structure so far unique to one strand of the satellite of tobacco ringspot virus (and a closely related virus). The HDV genome and antigenome cleavage structures do not seem to belong to either of these classes.

One puzzling aspect of the in vitro studies of KUO et al. (1988b) was that while a minimum sequence capable of undergoing self-cleavage could be defined, a slightly longer sequence was not functional. One such extra sequence was that which allowed the cleavage region to fold, intramolecularly, into part of the predicted unbranched rod structure. This inhibition could also be achieved intermolecularly, by the presence, in addition to a short molecule that was competent for cleavage, of a second RNA that allowed the formation by intermolecular base pairing, with this short RNA of a structure analogous to the unbranched rod structure. The data that the rod structure was inhibitory to self-cleavage suggested the corollary, that the rod structure might be an inducer of self-ligation, as discussed in Sect. 3.7.

3.7 Self-Ligation

SHARMEEN et al. (1989) sought to detect self-ligation on HDV RNA for three reasons. (1) There was already a precedent with the plant agents; (2) the model, introduced in Sect. 3.6, of the rod structure as an inducer of self-ligation seemed plausible; and (3) USHER and MCHALE (1976) had already described what proved to be another example of self-ligation. Their work had been concerned with obtaining protein-free ligation of hexamers of adenylic acid. These hexamers, like the cleavage products of HDV RNA, contained a 5′-hydroxyl and a 2′,3′-cyclic monophosphate. The hexamers could be induced to undergo efficient self-ligation by the presence of a backbone molecule, polyuridylic acid, that served to align the hexamers head-to-tail. Therefore, SHARMEEN et al. took two small fragments derived by self-cleavage of antigenomic HDV RNA, aligned these on an RNA from the other side of the rod structure, and using the conditions defined by USHER and MCHALE, obtained an efficient self-ligation reaction. One puzzling aspects of this work was that at most half of the self-ligated molecules could undergo a subsequent cleavage. It is possible that the other half had ligated not by a 3′–5′ linkage but by a nonbiological 2′–5′ linkage.

WU and LAI (1989) have also observed a self-ligation. They found that as long as the two self-cleavage products were allowed to remain associated after cleavage then ligation could be achieved simply by the removal of the magnesium ions.

3.8 RNA-Directed RNA Synthesis

As discussed above, the available data suggest that HDV, just like the plant agents, is replicated by RNA-directed RNA synthesis. Two important questions follow: where does this synthesis take place, and how is it achieved? As implied in Sect. 3.2, there is the suggestion that the location is the nucleolus, consistent with the findings for one of the plant viroids (HARDERS et al. 1989). The polymerase(s) involved has yet to be identified. It probably is not virus-coded, considering that the delta antigen is, so far, the only known viral protein and, at 22 kDa, is probably too small to act alone as an RNA polymerase. Thus, viral RNA synthesis probably requires a host polymerase. As known host polymerases normally use DNA rather than RNA templates, there will have to be a redirection. As reviewed by SYMONS et al. (1985), there are many claims for RNA polymerases which can be redirected in vitro to copy viroid RNAs. However, as SYMONS et al. point out, the transcription must first be understood in a more biological setting. There is only one report of experiments on viroid transcription in isolated nuclei; SPIESMACHER et al. (1985) claimed that DNA-dependent RNA polymerases I and II were used for the synthesis of genomic and antigenomic RNAs, respectively. Currently, such nuclear run-on assays are being applied to HDV in an attempt to determine which polymerase is actually redirected.

An important related question is the location on the HDV genome and antigenome of the sites of initiation of RNA-directed RNA synthesis. One preliminary piece of information is that the antigenomic RNA that acts as mRNA for the delta antigen (Sect. 4.3) appears to start 5 bases away from the end of the intramolecular rod structure (HSIEH et al. 1990). We do not know whether initiation of the opposite strand, the genomic RNA, occurs at one or multiple sites.

Yet another problem is to determine the *cis*-acting sequence features that are used to specify the initiation of genomic and antigenomic RNA transcription. Maybe the most direct way to solve this will be from a study of the sequences of HDV minivariants (Sect. 6) that are replication competent. It is also possible that a simple examination of the sequences conserved on such minivariants might help us to deduce the sequences essential for initiation of RNA synthesis.

4 The Delta Antigen

4.1 Structure

Three laboratories have published the complete sequence of the HDV genome (Sect. 2.2). From these studies, two different lengths for the open reading frame of the delta antigen have been predicted. As represented in Fig. 1, the amino terminus is shared, but the carboxy terminus is such that the overall length is either 195 or 214 amino acids. At this time there is an incompleteness in the published experiments for these two forms, which makes it difficult both in this and subsequent sections to comment precisely about the structure and function of the two forms. As discussed in Sect. 4.4, only the 195-aa form will support genome replication.

Immunoblot analyses of liver and serum samples from HDV-infected animals usually reveal two electrophoretic forms of the delta antigen (BERGMANN and GERIN 1986; POHL et al. 1987; BONINO et al. 1986; WEINER et al. 1988). Different laboratories have deduced different MWs; here we will use 22 and 24 kDa, since these values are closest to those predicted from the nucleotide sequence. Using cDNA it has been possible to express the two forms, one at a time, not only in animal cells but also in yeast and bacteria. The small form terminates with an amber codon, so that in appropriate bacteria, partial suppression leads to the synthesis of the two forms from a single RNA sequence (WEINER et al. 1988).

From the predicted sequences of the delta antigen two potentially important features have been noted. CHAO et al. (1990) have noted a sequence strikingly similar to the proven nuclear localization signal of the glucocorticoid receptor protein. LAI et al. (1990) have noted the presence of leucines ordered, approximately, in what LANDSCHULZ e al. (1989) have called a leucine zipper, a structure that is implicated as a facilitator of protein-protein interactions.

The large form of delta antigen has been shown by CHANG et al. (1988) to be modified in vivo by phosphorylation at serines, although the number of modifications is not known. The small form is more highly basic, with a predicted charge (excluding posttranslational modifications) of +12.

Both the large and small forms of delta antigen have been demonstrated to act as RNA binding proteins (CHANG et al. 1988; LAI et al. 1990; MACNAUGHTON et al. 1990). It is claimed that this binding has significant specificity for HDV-related RNAs, both genomic and antigenomic. To obtain more details CHANG et al. (1988) have made recombinant proteins expressing various portions of the large protein. They found that binding is primarily via a central domain, but there is also an advantage for the additional presence of the amino-terminal domain.

LAI et al. (1990) have also claimed that in vitro the delta antigen can make an apparently covalent interaction with HDV genomic RNA. They used as a substrate a 98-b genomic RNA fragment that was considered to contain the catalytic site for self-cleavage. They found that other regions of the genomic RNA or the antigenomic RNA failed to make this interaction. These binding studies were, however, carried out under nonphysiological conditions, and their significance is not yet clear.

4.2 Function

Both the large and small forms of the delta antigen can be found in virions; this could be interpreted as evidence that they both can function in the packaging of HDV RNA. This packaging is certainly consistent with the RNA binding data discussed in the previous section. One still needs to explain the strand-specificity of packaging, which was not seen in the binding studies.

The delta antigen may, of course, have functions in addition to regulating RNA packaging. We know, as discussed in Sect. 4.4, that the small form is essential for genome replication. In this role the large form will not function and may even be a suppressor of the function of the small form. What then is the role of the small form in genome replication? CHAO et al. (1991) have suggested that at least one function(s) may be provided during the initiation of infection, by the antigen present within the virions, since delta RNA alone is incapable of initiating a round of replication (GLENN et al. 1990). An obvious candidate role for the small antigen would be as a facilitator of RNA-directed RNA transcription. This may soon be resolved by studies with cell-free extracts, as suggested in Sect. 3.1.6. Another role may be in the synthesis of mRNA for the delta antigen, as discussed in Sect. 4.3, or in the regulation of that synthesis, as discussed in Sect. 5.

4.3 mRNA

Probably the major feature that makes HDV stand out from the subviral agents of plants is the existence of the delta antigen. Only some of the plant agents code for a protein, and for those that do, the protein has only a structural role, unlike the delta antigen which is not only a structural protein but is also essential for the replication of the genome. Recently, HSIEH et al. (1990) have resolved some of the questions as to how the delta antigen is translated. They have shown that in infected cells and transfected cells there exists a polyadenylated RNA that is about 800 b in length. The same species has been reported by JILBERT et al. (1991). This RNA is antigenomic, and its 5′ and 3′ ends have been mapped. It spans the open reading frame for both the small and large forms of the delta antigen and is found in the cytoplasm of the infected cell. These data are consistent with the RNA being the mRNA for the translation of delta antigen.

In animal cells the polyadenylation of mRNAs is in part directed by a signal, AAUAAA, just 5′ of the adenylation site (BIRNSTIEL and SCHAUFELE 1989). The HDV genome has no such signal, while the antigenome has but one, and its location is as expected for the processing of the observed mRNA. HSIEH et al. used site-directed mutagenesis to change this signal to UUUAAA. The resultant HDV mutant, when used in a DNA transfection, neither made delta antigen nor was able to initiate HDV RNA synthesis. The block could be relieved by cotransfection with a second plasmid that expressed only the delta antigen. Thus, the poly(A) signal is essential for the expression of the delta antigen.

One interesting feature of the delta antigen mRNA is that it is present in relatively low amounts. The amount of the putative mRNA is usually at least 500 times lower than the amount of genomic RNA. The control of the synthesis of this RNA is discussed further in Sect. 5.

At this time there is no positive evidence that the putative mRNA can be translated in vitro, but lack of success to date may be due to the relatively small amounts that have been available or to inhibition by contaminating genomic RNA, which effectively acts as antisense RNA (WEINTRAUB 1990).

4.4 Variation

As mentioned earlier, three complete HDV sequences have been published (WANG et al. 1986, 1987; MAKINO et al. 1987; KUO et al. 1988a) as well as additional partial sequences (KOS et al. 1986; SALDANHA et al. 1987). Two species of delta antigen has been predicted: These are 195 and 214 aa in length, with a common amino terminus and a difference of 19 amino acids at the carboxy terminus. WANG et al. 1986, 1987) and more recently LAI et al. (1990) have seen such sequence heterogeneity in the RNA from a single patient. A clue to the origin of these two forms of the delta antigen comes from the work of SUREAU et al. (1989) and LUO et al. (1990), to be discussed in Sect. 6.

5 Regulation of mRNA Transcription Relative to Genome Replication

In Sect. 4.3, it was noted that the synthesis of the delta antigen is dependent upon a specific cytoplasmic mRNA. Since the replication of unit length genomic RNA takes place in the nucleus and both species are needed, this means that there has to be what we will define here as "regulation." This regulation is represented diagrammatically in Fig. 2. As part of this discussion we will include the observation that the amount of the cytoplasmic mRNA is at least 500 times lower than that of the unit length genomic RNA and 50 times lower than unit length antigenomic RNA in the nucleus. How then is this regulation achieved? HSIEH et al. (1990) have speculated that the regulation is achieved both at the level of RNA synthesis and at the level of mRNA stability. Consider, first, RNA synthesis and the concepts that might be borrowed from the negative-strand viruses. HDV is not a true virus, but it is negative-stranded; thus, like these viruses, it has to regulate the amount of RNA for messenger versus the amount for genome replication. The negative-strand viruses are known to do this at the level of RNA

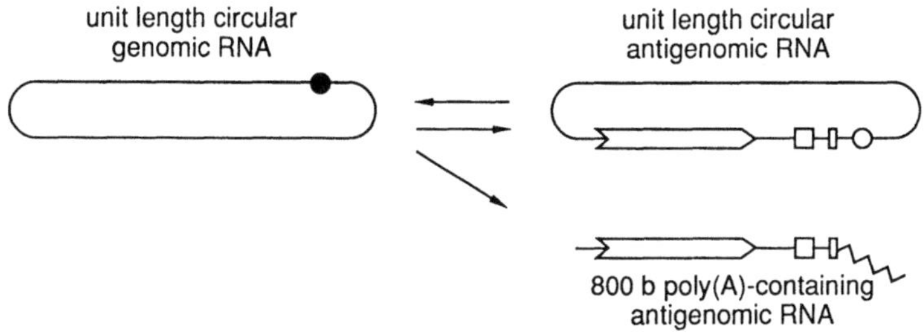

Fig. 2. Representation of the regulation of HDV genome replication relative to the synthesis of mRNA for the delta antigen. The genome and antigenome are drawn as elongated circles to express both the circularity and the ability to fold into an unbranched rod structure. The self-cleavage sites on the genome and antigenome are indicated as *closed* and *open circles*, respectively. The antigenome also shows the coding region for the delta antigen (*open arrow*), the polyadenylation signal (*open square*), and adenylation site (*open rectangle*). On the mRNA, the poly(A) is indicated by the *wavy line*

transcription, and the regulation is known to be directed by virus-coded proteins (BANERJEE 1987; KRUG et al. 1987). Thus, HDV may also do this, and with only one HDV-coded protein to choose from, the delta antigen may carry out such a role. What about regulation at a second level, stability of the mRNA? It this were true it would still have to be secondary, that is, in addition to a primary regulation at the level of RNA synthesis. The second level is introduced to help explain why the relative level of mRNA is so low.

Another model of regulation has also been proposed (TAYLOR et al. 1989). The hypothesis is that antigenomic RNA synthesis is initiated from a single location to form a rolling-circle, with the self-cleavage reaction cutting out unit length molecules. When these transcripts first become long enough to include the polyadenylation signals and maybe also the self-cleavage site, a choice between the two has to be made. However, independent of this choice, the replication complex in the absence of reinitiation will not have another opportunity to make an 800-b mRNA; all future transcripts from it will lead to unit length antigenomic RNAs when there is a suppression of the poly(A) addition signals.

6 Variant Genomes

For HDV, as for other viruses, replication can lead to sequence variants, some of which are still able to replicate. Consider first those variant HDV genomes that are about the same length as wild-type reference stocks but with a limited number of sequence changes. The existence of such variants became clear following cDNA cloning and sequencing of HDV. In the studies of WANG et al. (1986, 1987), it was found that viruses in the serum of a single infected animal contained RNA sequence variants. This is, of course, true for most other viruses, including the plant agents (KEESE et al. 1988). With the sequencing of other HDV isolates (MAKINO et al. 1987), it became clear that there were not only base substitutions but also insertions and deletions relative to the HDVs sequenced by WANG et al. (1986). As mentioned earlier (Sect. 1), there are HDV outbreaks that produce unusually severe liver disease, and from the sequence of the HDV RNA obtained from cases, it may be possible to recognize genome feature(s) essential for an increase in pathogenicity. Such studies have been attempted with the plant viroids, as reviewed by KEESE et al. (1988), and there are precedents for as little as a single base change being a determinant of pathogenicity. For both the plant agents and for HDV it is quite possible that different, more or less pathogenic strains, arise in several ways.

Another kind of HDV variant has recently been observed in the work of LUO et al. (1990) and SUREAU et al. (1989). This is a variant that arises both with high efficiency and specificity during HDV genome replication within a single cell, in the absence of further cell-to-cell spread. It was observed that when cultured cells were transfected with a cDNA clone of HDV, there was not only the initiation of HDV genome replication but also the accumulation of genomes with a specific sequence alteration. This alteration also occurred in a chimpanzee transfected by HDV cDNA. The alteration was in the amber termination codon for the 195-aa delta antigen, thereby directing the coding of the amino acid tryptophan and creating a delta antigen that was 19 aa longer, for a total of 214 aa, i.e., the large delta antigen. This modification only occurred in transfection experiments when there was RNA-dependent RNA synthesis, and in fact, the extent of modification depended on the extent of genome synthesis. LUO et al. have speculated that this change might involve RNA unwindase activity. This is a deaminase activity which is present in most animal cells (BASS and WEINTRAUB 1988; WAGNER and NISHIKURA 1988), and has recently been implicated in the modification of the genomes of negative-strand viruses (BASS et al. 1989) and in specific developmental control of mRNA species (KIMELMAN and KIRSCHNER 1989).

LUO et al. also showed by transfection experiments that the large form of the delta antigen produced as a result of this sequence alteration, unlike the small form, was unable to support HDV genome replication. They thus speculated that the larger form might in some way reduce the extent of HDV replication, leading in the short term to a milder acute disease or in the long term to an increased chance of progressing to chronicity.

Yet another class of HDV variants are those with genomes that are significantly smaller than the normal size of about 1700 b. These are called minivariants (TAYLOR et al. 1990a). Such species, with lengths down to about 400 b, have been detected in the livers of infected chimpanzees and especially infected woodchucks. Apparently tne molecules lack the *cis*-acting sequences needed for packaging, because they are not found in virions. However, the observation that they are produced in relatively significant amounts implies that they must have the *cis*-acting elements required for replication. Thus, it will be informative to isolate and sequence these species.

7 Interaction with the Hepadnavirus

Of course, the major interaction of HDV with HBV occurs during packaging. Presumably there are *cis*-cting signals both on the HDV RNAs and transacting signals on the HBV proteins and*or on the delta antigen that regulate this interaction. At present, however, nothing is known about the regulating signals.

Earlier (Sect. 3.5), it was shown that HDV RNA synthesis does not depend upon the hepadnavirus. However, as summarized below, in the biological setting of an in vivo infection, the situation is still quite complex:

1. The hepadnavirus does provide the coat, but it is not yet clear whether this coat is exactly the same as present on the hepadnaviruses. Some data of BONINO et al. (1986) argue for a difference in the relative amounts of the various related forms of HBV surface antigen in HDV as compared with HBV.
2. There are data that HDV infection can, at least transiently, have a suppressive effect on hepadnavirus replication (e.g., CHEN et al. 1988 and SUREAU et al. 1989).
3. From what we now know, a significant part of an in vivo HDV infection may be nonproductive, that is, resulting from the spread of HDV particles to liver cells (and perhaps to nonliver cells) that are not already infected by HBV nor likely to be infected. Such a situation might occur, for example, following HDV infection of a surface antigen carrier that is no longer viremic. There are actual immunofluorescence microscopy and in situ hybridization data from GOWANS et al. (1988) to prove such events occur in natural infections. Such nonproductive infection may in theory have significant overall consequences for the host. These consequences may be deleterious or as speculated below may in some situations even be advantageous.
4. A number of investigators, working with HDV-infected patients experiencing progressive liver failure, have as a last resort offered a liver transplantation to their patients. It has been noted that such patients can promptly establish HDV RNA synthesis in the new liver (as distinct from a full replication cycle with packaging of new virions). The spread of delta to the new liver seems in turn to give this organ enhanced resistance to infection by HBV, which in these studies was also being produced prior to the transplantation (GRIDELLI et al. 1991). This could be the explanation for the better prognosis for a liver transplant when the patient has a prior HDV infection, relative to patients infected with HBV alone.

8 Future Directions

Throughout this review there has been an attempt to discuss questions regarding HDV and its replication that are currently accessible to experimental analysis. Prominent examples include the mechanism of RNA-directed RNA synthesis, the development of an in vivo system for genome replication, the regulation of the mRNA for the delta antigen, and the roles of the delta antigen. Another important problem is the mechanism of packaging. We have suggested that there might be *cis*-acting sequences on the HDV genome, but at the same time, it might be possible that other viruses could provide the proteins that would allow for packaging of HDV into infectious particles. It is possible to speculate that other subviral agents analogous to HDV will be found to be associated with human diseases (TAYLOR 1990 b).

A particularly important direction for the future will be to apply the information we have obtained about HDV replication to problems of human disease. For example, we have with HDV the wherewithal to deliver an RNA that can, with minimal help, be amplified 300 000-fold in a single cell. If the HDV genome were modified to contain antisense (WEINTRAUB 1990) or antigene (HAZELOFF and GERLACH 1988) activities, one might achieve the ability to inactivate specific genes efficiently. With an appropriate delivery system, one might be able to ameliorate certain chronic infections, such as HBV, which is normally a major risk factor for the development of hepatocellular carcinoma (BEASLEY 1988).

Acknowledgments. Financial support was from grants CA-06927, RR-05539, and AI-26522 from the National Institutes of Health, grant MV-7M from the American Cancer Society, and an appropriation from the Commonwealth of Pennsylvania. Discussions of unpublished work were with Mei Chao, Sen-Yung Hsieh, Guangxiang Luo, and Wang-Shick Ryu as well as with Jeff Glenn. Valuable comments on the manuscript were given by Bill Mason, Allison Jilbert, and Tom London.

References

Banerjee AK (1987) Transcription and replication of rhabdoviruses. Bacteriol Rev 55: 599–629

Bass BL, Weintraub H (1988) An unwinding activity that covalently modifies its double-stranded RNA substrate. Cell 55: 1089–1098

Bass BL, Weintraub H, Cattaneo R, Billeter MA (1989) Biased hypermutation of viral RNA genomes could be due to unwinding/modification of double-stranded RNA. Cell 56: 331

Beasley RP (1988) Hepatitis B virus: the major etiology of hepatocellular carcinoma. In: Fortner JG, Rhoads JE (eds) Accomplishments in cancer research: 1987 prize year. Lippincott, Philadelphia pp 80–106

Belinsky M, Dinter-Gottlieb G (1991) Characterizing the self-cleavage of a 135 nucleotide ribozyme from genomic hepatitis delta virus. In: Gerin JL, Purcell RH, Rizzeto M (eds) The hepatitis delta virus. Liss, New York, pp. 265–274

Bergmann KF, Gerin JL (1986) Antigens of hepatitis delta virus in the liver and serum of humans and animals. J Infect Dis 154: 702–705

Birnstiel ML, Schaufele FJ (1989) Structure and function of minor snRNPs. In: Birnstie ML (ed) Small ribonucleoprotein particles. Springer, Berlin Heidelberg New York, pp 155–182

Bonino F, Hoyer W, Shih JW-K, Rizzetto M, Purcell RH, Gerin JL (1984) Delta hepatitis agent: structural and antigenic properties of the delta-associated particles. Infect Immun 43: 1000–1005

Bonino F, Heermann K-H, Rizzetto M, Gerlich WH (1986) Hepatitis delta virus: protein composition of delta antigen and its hepatitis B virus-derived envelope. J Virol 58: 945–950

Bonino F, Negro F, Baldi M, Brunetto M, Chiaberge E, Capalbo M, Maran E, Lavarini C, Rocca N, Rocca G (1987) The natural history of chronic delta hepatitis. In: Rizzetto M, Gerin JL, Purcell RH (eds). The hepatitis delta virus and its infection. Liss New York, pp 145–152

Bonino F, Negro F, Brunetto MR, Verme G (1990) Hepatitis delta infection. Prog Liver Dis 9: 485–496

Branch AD, Robertson HD (1984) A replication cycle for viroids and small infectious RNAs. Science 223: 450–455

Branch AD, Levine BJ, Baroudy BM, Buckler-White A, Gerin JL, Robertson HD (1991) The novel tertiary structure in delta RNA may function as a ribozyme control element. In: Gerin JL, Purcell RH, Rizzeto M (eds) The hepatitis delta virus. Liss, New York, pp. 257–264

Brown F, Martin SJ (1965) A new model for virus ribonucleic acid replication. Nature 208: 861–863

Buitrago B, Popper H, Hadler SC, Thung SN, Gerber MA, Purcell RH, Maynard JE (1986) Specific histologic features of Santa Marta hepatitis: a severe form of hepatitis δ-virus infection in northern South America. Hepatology 6: 1285–1291

Buzayun JM, Gerlach WL, Bruening G (1986) Non-enzymatic cleavage and ligation of RNAs complementary to a plant virus satellite RNA. Nature 323: 349–353

Chang M-F, Baker SC, Soe LH, Kamahora T, Keck JG, Makino S, Govindarajan S, Lai MMC (1988) Human hepatitis delta antigen is a nuclear phosphoprotein with RNA binding activity. J Virol 62: 2403–2410

Chao M, Hsieh S-Y, Luo G, Taylor J (1991) The antigen of human hepatitis delta virus: the significance of the two major electrophoretic forms. In: Gerin JL, Purcell RH, Rizzeto M (eds) The hepatitis delta virus. Liss, New York, pp. 257–281

Chen P-J, Kalpana G, Goldberg J, Mason W, Werner B, Gerin J, Taylor J (1986) The structure and replication of the genome of the hepatitis delta virus. Proc Natl Acad Sci USA 83: 8774–8778

Chen P-J, Chen D-S, Chen C-R, Chen Y-Y, Chen H-MH, Lai M-Y, Sung J-L (1988) δ infection in asymtomatic carriers of hepatitis B surface antigen: low prevalence of δ activity and effective suppression of hepatitis B virus replication. Hepatology 8: 1121–1124

Choi S-S, Rasshofer R, Roggendorf M (1988) Propagation of woodchuck hepatitis delta virus in primary woodchuck hepatocytes. Virology 167: 451–457

Choi S-S, Rasshofer R, Roggendorf M (1989) Inhibition of hepatitis delta virus RNA replication in primary woodchuck hepatocytes. Antiviral Res 12: 213–222

Davis GL, Balart LA, Schiff ER, Lindsay K, Bodenheimer HC, Prrillo RP, Carey W, Jacobson IM, Payne J, Dienstag JL, Van Thiel DH, Tamburro C, Lefkowitch J, Albrecht J, Meschievitz C, Ortego TJ, Gibas A, and the Hepatitis International Therapy Group (1989) Treatment of chronic hepatitis C with recombinant interferon alpha: a multicenter randomized, controlled trial. N Eng J Med 321: 1501–1510

Diener TO (1989) Circular RNAS: relics of precellular evolution? Proc Natl Acad Sci USA 86: 9370–9374

Diener TO, Prusiner SB (1985) The recognition of subviral pathogens. In: Marmorosch K, McKelvey JJ Jr (eds) Subviral pathogens of plants and animals: viroids and prions. Academic, Orlando, pp 3–20

Dinter-Gottlieb G (1986) Viroids and virusoids are related to group I introns. Proc Natl Acad Sci USA 83: 6250–6254

Ellens H, Doxey S, Glenn JS, White JM (1989) Delivery of macromolecules into cells expressing a viral membrane fusion protein. Methods Cell Biol 31: 155–176

Field HJ, Wildy P (1978) The pathogenicity of thymidine kinase-deficient mutants of herpes simplex virus in mice. J Hyg (Lond) 81: 267–277

Forster AJ, Symons RH (1987) Self-cleavage of viroid RNA is performed by the proposed 55-nucleotide active site. Cell 50: 9–16

Ganem D, Varmus HE (1987) The molecular biology of hepatitis B viruses. Annu Rev Biochem 56: 651–693

Glenn JS, Taylor JM, White JM (1990) In vitro synthesized hepatitis delta virus RNA initiates genome replication in cultured cells. J Virol 64: 3104–3107

Gowans EJ, Baroudy BM, Negro F, Ponzetto A, Purcell RH, Gerin JL (1988) Evidence for the replication of hepatitis delta virus RNA in nuclei after in vivo infection. Virology 167: 274–278

Gridelli B, Grendele M, Colledan M, Rossi G, Fassati LR, Feria G, Maggi U, Reggiani P, Doglia M, Lunghi G, Galmarini D (1991) Liver transplantation (LTx) and delta virus (DV) infection. In: Gerin JL, Purcell RH, Rizzeto M (eds) The Hepatitis Delta Virus, Liss, New York, p. 471

Haas B, Klanner A, Ramm K, Sänger HL (1989) The 7S RNA from tomato leaf tissue resembles a signal recognition particle RNA and exhibits a remarkable sequence complementarity to viroids. EMBO J 7: 4063–4074

Hadidi A (1986) Relationships of viroids and certain other plant pathogenic nucleic acids to group I and II introns. Plant Mol Biol 7: 129–142

Harders J, Lukàcs N, Robert-Nicoud M, Jovin TM, Riesner D (1989) Imaging of viroids in nuclei from tomato leaf tissue by in situ hybridization and confocal laser scanning microscopy. EMBO J 8: 3941–3949

Hazeloff J, Gerlach WL (1988) Simple RNA enzymes with new and highly specific endoribonuclease activities. Nature 334: 585–591

He L-F, Ford E, Purcell RH, London WT, Phillips J, Gerin JL (1989) The size of the hepatitis delta agent. J Med Virol 27: 31–33

Hoofnagle JH (1989) Type D (Delta) hepatitis. J A M A 261: 1321–1325

Hutchins CJ, Rathjen PD, Forster AC, Symons RH (1986) Self-cleavage of plus and minus RNA transcripts of avocado sunblotch viroid. Nucleic Acids Res 14: 3627–3640

Hsieh S-Y, Chao M, Coates L, Taylor J (1991) Hepatitis delta virus genome replication: a polyadenylated mRNA for delta antigen. J Virol 64: 3192–3198

Jilbert AR, Gowans EJ, Macnaughton TB, Burrell CJ (1991) Characterization of a subgenomic HDV-specific poly(A) + RNA species isolated from human hepatoma cells supporting HDV RNA replication. In: Gerin JL, Purcell RH, Rizzeto M (eds) The hepatitis delta virus. Liss, New York, pp. 309–314

Keese P, Visvader JE, Symons RH (1988) Sequence variability in plant viroid RNAs. In: Domingo E, Holland JJ, Ahlquist P (eds) RNA genetics, vol III, variability of RNA genomes. CRC Press, Boca Raton, pp 71–98

Kimelman D, Kirschner MW (1989) An antisense mRNA directs the covalent modification of the transcript encoding fibroblast growth factor in Xenopus oocytes. Cell 59: 687–696

Kos A, Dijkema R, Arnberg AC, Van der Meide PH, Schelleken H (1986) The hepatitis delta virus possesses a circular RNA. Nature 322: 558–560

Krug RM, St Angelo C, Broni B, Shapiro G (1987). Transcription and replication of influenza virion RNA in the nucleus of infected cells. Cold Spring Harbor Symp Quant Biol 52: 353–358

Kuo MY-P, Goldberg J, Coates L, Mason W, Gerin J, Taylor J (1988a) Molecular cloning of hepatitis delta virus RNA from an infected woodchuck liver: sequence, structure and applications. J Virol 62: 1855–1861

Kuo MY-P, Sharmeen L, Dinter-Gottlieb G, Taylor J (1988b) Characterization of self-cleaving RNA sequences on the genome and antigenome of human hepatitis delta virus. J Virol 62: 4439–4444

Kuo MY-P, Chao M, Taylor J (1989) Initiation of replication of the human hepatitis delta virus genome from cloned DNA: role of the delta antigen. J Virol 63: 1945–1950

Lai MMC, Chao Y-C, Chang M-F, Lin J-H, Gust I (1991) Functional studies of hepatitis delta antigen and delta virus RNA. In: Gerin JL, Purcell RH, Rizzeto M (eds) The hepatitis delta virus. Liss, New York, pp. 283–292

Lanschulz WH, Johnson PF, McKnight (1989) The leucine zipper: a hypothetical structure common to a new class of DNA binding proteins. Science 240: 1759–1764

Luo G, Chao M, Hsieh S-Y, Sureau C, Nishikura K, Taylor J (1990) A specific base transition occurs on replicating hepatitis delta virus RNA. J Virol 64: 1021–1027

Macnaughton TB, Gowans EJ, Reinboth B, Jilbert AR, Burrell CJ (1990) Stable expression of the small polypeptide of hepatitis delta virus antigen in a eukaryotic cell line. J Gen Virol 71: 1339–1345

Makino S, Chang M-F, Shieh C-K, Kamahora T, Vannier DV, Govindarajan S, Lai MMC (1987) Molecular cloning and sequencing of a human hepatitis delta virus RNA. Nature 329: 343–346

Negro F, Bergmann KF, Baroudy BM, Satterfield WC, Popper H, Purcell RH, Geril JL (1988) Chronic hepatitis D virus (HDV) infection in hepatitis B virus carrier chimpanzees experimentally superinfected with HDV. J Infect Dis 158: 151–159

Negro F, Gerin JL, Purcell RH, Miller RH (1989) Basis of hepatitis delta virus disease. Nature 341: 111

Petcu DJ, Aldrich CE, Coates L, Taylor JM, Mason WS (1988) Suramin effectively inhibits in vitro infection by duck hepatitis B virus, Rous sarcoma virus and hepatitis delta virus. Virology 167: 385–392

Pohl C, Baroudy BM, Bergmann KF, Cote PJ, Purcell RH, Hoofnagle J, Gerin JL (1987) A human monoclonal antibody that recognizes viral polypeptides ans in vitro translation products of the genome of the hepatitis delta virus. J Infect Dis 156: 622–629

Ponzetto A, Cote PJ, Popper H, Boyer BH, London WT, Ford EC, Bonino F, Purcell RH, Gerin JL (1984) Transmission of the hepatitis B-associated delta agent to the eastern woodchuck. Proc Natl Acad Sci USA 81: 2208–221

Ponzetto A, Rapicetta M, Forzani B, Smedile A, Hele C, Morace G, Di Rienzo A, Palladino A, Gerin JL (1987) Hepatitis delta virus infection in pekin ducks chronically infected by the duck hepatitis B virus. In: Rizetto M, Gerin JL, Purcell RH (eds) The hepatitis delta virus and its infection. Liss, New York, pp 47–49

Popper H, Thung SN, Gerber MA, Hadler SC, De Monzon M, Ponzetto A, Anzola E, Rivera D, Mondolfi

A, Bracho A, Francis DP, Gerin JL, Maynard JE, Purcell RH (1983) Histological studies of severe delta agent infection in Venezuelan Indians. Hepatology 3: 906–912
Prody GA, Bakos JT, Buzayan JM, Schneider IR, Bruening G (1986) Autolytic processing of dimeric plant virus satellite RNA. Science 231: 1577–1580
Purcell RH, Satterfield WC, Bergmann KF, Smedile A, Ponzetto A, Gerin JL (1987) Experimental hepatitis delta virus infection in the chimpanzee. In: Rizzetto M, Gerin JL, Purcell RH (eds) Hepatitis delta virus and its infection. Liss, New York, pp 27–36
Rizzetto M (1983) The delta agent. Hepatology 3: 729–737
Rizzetto M, Canese MG, Arico J, Crivelli O, Bonino F, Trepo CG, Verme G (1977) Immunofluorescence detection of a new antigen-antibody system (δ-antiδ) associated to the hepatitis B virus in the liver and in the serum of HBsAg carriers. Gut 18: 997–1003
Rizzetto M, Hoyer B, Canese MG, Shih JW-K, Purcell RH, Gerin JL (1980) Delta agent: association of delta antigen with hepatitis B surface antigen and RNA in serum of delta-infected chimpanzees. Proc Natl Acad Sci USA 77: 6124–6128
Saldanha JA, Thomas HC, Monjardino JP (1987) Cloning and characterization of a delta virus cDNA sequence derived from a human source. J Med Virol 22: 323–331
Sänger HL (1987) Viroid replication. In: Diener TO (ed) The viroids. Plenum, New York, pp 117–166
Schödel F, Sprengel R, Weimer T, Fernholz D, Schneider R, Will H (1989) Animal hepatitis B viruses. Adv Viral Oncol 8: 73–102
Sharmeen L, Kuo MY-P, Dinter-Gottlieb G, Taylor J (1988) The antigenomic RNA of human hepatitis delta virus can undergo self-cleavage. J Virol 62: 2674–2679
Sharmeen L, Kuo MY-P, Taylor J (1989) Self-ligating sequences on the antigenome of human hepatitis delta virus. J Virol 63: 1428–1430
Smith CWJ, Porro EB, Patton JG, Nadal-Ginard B (1989) Scanning from an independently specified branch point defines the 3′ splice site of mammalian introns. Nature 342: 243–247
Spiesmacher E, Mühlbach H-P, Tabler M, Sänger HL (1985) Synthesis of (+) and (−) RNA molecules of potato spindle tuber viroid (PSTV) in isolated nuclei and its impairment by transcription inhibitors. Biosci Rep 5: 251–265
Summers J, Mason WS (1982) Replication of the genome of a hepatitis B-like virus by reverse transcription of an RNA intermediate. Cell 29: 403–415
Sureau C, Taylor J, Chao M, Eichberg JE, Lanford RE (1989) Cloned hepatitis delta virus complementary DNA is infectious in the chimpanzee. J Virol 63: 4292–4297
Sureau C, Taylor J, Chao M, Eichberg JW, Lanford RE (1991) In vitro tissue culture system for the study of HDV infectivity. In: Gerin JL, Purcell RH, Rizzeto M (eds) The hepatitis delta virus. Liss, New York, p. 464
Symons RH (1989a) Pathogenesis by antisense. Nature 338: 542–543
Symons RH (1989b) Self-cleavage of RNA in the replication of small pathogens of plants and animals. TIBS 14: 445–450
Symons R, Haseloff J, Visvader JE, Keese P, Murphy PJ, Gill DS, Gordon KHJ, Bruening G (1985) On the mechanism of replication of viroids, virusoids, and satellite RNAs. In: Marmorosch K, McKelvey JJ Jr (eds) Subviral pathogens of plants and animals: viroids and prions. Academic, Orlando, pp 235–263
Taylor JM (1990a) Human hepatitis delta virus: structure and replication of the genome. CTMI (to be published)
Taylor JM (1990b) Structure and replication of hepatitis delta virus. Semin Virol 1: 135–141
Taylor JM, Kuo M, Chen P-J, Kalpana G, Goldberg J, Aldrich A, Coates L, Mason W, Summers J (1987a) Replication of hepatitis delta virus. In: Robinson W, Koike K, Will H (eds) Hepadna Viruses. Liss, New York, pp. 541–548
Taylor JM, Mason W, Summers J, Goldberg J, Aldrich C, Coates L, Gerin J, Gowans E (1987b) Replication of human hepatitis delta virus in primary cultures of woodchuck hepatocytes. J Virol 61: 2891–2895
Taylor JM, Sharmeen L, Kuo M, Dinter-Gottlieb G (1989) The self-cleaving RNAs of human hepatitis delta virus. In: Cech T (ed) Molecular biology of RNA. Liss, New York, pp 99–108 (ULCA symposia on molecular and cellular biology, new series, vol 94)
Taylor JM Chao M, Kuo M, Sharmeen L (1990a) The replication of hepatitis delta virus: prospects. In: IV International symposium on "progress in viral hepatitis", Madrid, January 26–27, 1989, Medscript, London, pp in press
Taylor JM, Chao M, Kuo M, Sharmeen L, Hsieh S-Y (1990b) Human hepatitis delta virus: unique or not unique. In: Proceedings of 2nd international symposium on positive strand viruses A.S.M. Publications, pp. 20–24
Uhlenbeck OC (1987) A small catalytic oligoribonucleotide. Nature 328: 596–600

Usher DA, McHale AH (1976) Nonenzymic joining of oligoadenylates on a polyuridylic acid template. Nature 328: 586–600
Wagner RW, Nishikura K (1988) Cell cycle expression of RNA duplex unwindase activity in cells. Mol Cell Biol 8: 770–777
Wang K-S, Choo Q-L, Weiner AJ, Ou J-H, Najarian RC, Thayer RM, Mullenbach GT, Denniston KJ, Gerin JL, Houghton M (1986) Structure, sequence and expression of the hepatitis delta viral genome. Nature 323: 508–513
Wang K-S, Choo Q-L, Weiner AJ, Ou J-H, Najarian RC, Thayer RM, Mullenbach GT, Denniston KJ, Gerin JL, Houghton M (1987) Corrigendum: structure, sequence and expression of the hepatitis delta viral genome. Nature 328: 456
Weiner AJ, Choo Q-L, Wang K-S, Govindarajan S, Redeker AG, Gerin JL, Houghton M (1988) A single antigenomic open reading frame of the hepatitis delta virus encodes the epitope(s) of both hepatitis delta antigen polypeptides $p24^{\delta}$ and $p27^{\delta}$. J Virol 62: 594–599
Weintraub HM (1990) Antisense RNA and DNA. Sci Am 40–46
Weissman C (1989) Single-strand RNA. Nature 337: 415
Wu H-N, Lai MMC (1989) Reversible cleavage and ligation of hepatitis delta virus RNA. Science 243: 652–654
Wu H-N, Lin Y-J, Lin F-P, Makino S, Chang M-F, Lai MMC (1989) Human hepatitis δ virus RNA subfragments contain an autocleavage activity. Proc Natl Acad Sci USA 86: 1831–1835
Young B, Hicke B (1990) Delta virus as a cleaver. Nature 343–28

Development of Antiviral Therapy for Chronic Infection with Hepatitis B Virus

P. L. MARION

1 Introduction

Infection with hepatitis B virus (HBV) can now be prevented through the use of several effective vaccines (MAYNARD et al. 1989). Yet for the millions of existing chronic carriers there is no effective treatment to clear the virus and thus limit the high risk of liver diseases associated with long-term infection; chronic

Division of Infectious Diseases, Department of Medicine, Stanford University School of Medicine, Stanford, CA 94305, USA hepatitis, cirrhosis, and hepatocellular carcinoma. A number of recent reviews have described the limited success of current antiviral therapies of patients chronically infected with HBV (ALEXANDER and WILLIAMS 1988; LEVER 1988; GARCIA and GENTRY 1989; DI BISCEGLIE and HOOFNAGLE 1989). The focus of this chapter will be on understanding why HBV is difficult to eradicate once chronic infection is established, on possible new targets for antiviral therapy in the life cycle of HBV, and on recently developed models for testing the efficacy of anti-hepadnaviral treatments.

2 Current Problems and New Approaches in Hepadnaviral Therapy

2.1 Limited Immune Response

Infection with HBV becomes chronic when a patient fails to clear the viremia within the first 6 months,. Development of chronic infection in neonates following perinatal exposure to carrier mothers is relatively common (BEASLEY and HWANG 1984) and is thought to be due to the immunotolerance of an immature immune system. Yet approximately 10% of newly infected adults also become chronic carriers, possibly because of an inadequate immune response in the early weeks and months of infection. Thus, the first problem associated with chronic infection by HBV is a limited or defective host response to viral antigens. Although this review will not focus on the development of immune modulators or stimulation of the immune system as therapy for HBV infection, this area is as important as that of arresting ongoing viral infection. Particularly encouraging in this domain is the recent observation that the titer of circulating HBV surface antigen in immunotolerant transgenic mice expressing this antigen can be lowered by immunization with a surface antigen of another subtype (C. POURCEL, personal communication). This finding opens the possibility of stimulating an immune response adequate to clear the viremia by vaccinating some of the existing carriers with heterologous subtypes.

2.2 HBV Is Not Cytopathic

A second problem in eliminating chronic HBV infection is that the infection does not usually kill the host cells. (See chapter by F.V. CHISARI, this volume and below for a discussion of chronic cytopathogenicity associated with viral gene expression.) Rather, the inflammation of the liver associated with viral infection appears to be primarily the result of the cellular immune response to viral antigens. One cannot resolve HBV infection by simply preventing de novo infection and allowing the already infected cells to die.

2.3 CCC DNA, a Relatively Invulnerable Form of HBV DNA

As described in preceding chapters, the hepadnaviral scheme of replication is distinct from that of other virus families. Incoming relaxed circular virion DNA is converted to a covalently closed, circular (CCC) DNA form residing in the nucleus of the hepatocyte. RNA transcribed from this DNA species is used to generate both viral antigens and the mature virion DNA via reverse transcription. The CCC DNA does not self-replicate and is thus unaffected by treatment with drugs inhibiting nucleic acid synthesis, such as adenine arabinoside, acyclovir, or phosphonoformate. However, these drugs are effective in inhibiting the formation of mature virion DNA (SHERKER et al. 1986; DI BISCEGLIE and HOOFNAGLE 1989). When treatment with these nucleic acid synthesis inhibitors is stopped, HBV synthesis is resumed via a continued generation of viral mRNA from the still-present nuclear CCC template. The problem of CCC resistance to antihepadnaviral drugs in patients and in model systems is compounded by the fact that the number of CCC DNA molecules per cell appears to be amplified during the viral life cycle by delivery to the nucleus of the mature virion DNA within completed cores, followed by conversion of the new viral DNA to the CCC DNA form (TUTTLEMAN et al. 1986a). Reinfection by exported virus is not necessary for the amplification (WU et al. 1990). This CCC DNA form is maintained at approximately 50 copies per cell in fully infected liver tissue (TUTTLEMAN et al. 1986a). It is not known how rapidly CCC DNA decays if viral replication and reinfection are fully arrested. CCC DNA is unlikely to be diminished in the liver by dilution due to cell replication, as hepatocytes in highly infected liver with little inflammation do not divide.

2.4 HBV Is Found in Nonhepatic Tissues

Even if therapy might eliminate HBV infection from hepatocytes, there may be subsequent reinfection due to a reservoir of HBV in other tissues (reviewed by MARION 1988). It has been demonstrated by several groups that hepadnaviral DNA is found in cells of the immune system, usually in a nonreplicative form. Mitogenic stimulation with lipopolysaccharide of woodchuck hepatitis virus (WHV)-infected woodchuck peripheral blood lymphocytes will induce viral replication in these cells bearing nonreplicative viral DNA (KORBA et al. 1989). Hence, any antiviral therapy in which the virus in nonhepatic tissues is unaffected or in which neutralizing antibody is not induced, may eventually fail to cure the patient.

2.5 Integration of HBV DNA into Liver Chromosomes

Although integration of viral DNA is not part of the basic life cycle of hepadnaviruses, viral DNA integration does occur during chronic infection and is readily detected by Southern blotting in many hepadnavirus-associated hepatocellular carcinomas (see chapter by C.E. ROGLER, this volume). In long-term carriers, integrated HBV DNA is frequently the only viral DNA form present in the liver. The integrated viral DNA is most commonly rearranged, with inversions and deletions. The genome is interrupted in such a way that generation of the complete pregenome needed to make infectious virus is not possible (NAGAYA et al. 1987). The S open reading frame (ORF) is often intact, and many of the individuals with integrated viral DNA make defective virus particles with only the envelope. It is not known whether the integrated viral DNA may occasionally be rearranged to yield infectious virus. The gene products of the integrated viral DNA may be associated with pathogenesis. Synthesis of the large surface antigen molecules bearing the preS1 region without synthesis of an adequate amount of the smaller surface antigen protein results in an intracellular accumulation of the preS1 protein accompanied by hepatocellular injury, eventually leading to hepatocellular carcinoma in transgenic mice (chapter by F.V. CHISARI, this volume; CHISARI et al. 1987, 1988). However, the synthesis of the smallest form of surface antigen alone is not known to be cytopathic. For individuals at the stage of chronic infection in which only defective particles are produced, drugs affecting replication of nucleic acid or nucleocapsid synthesis would not be helpful. Any therapy designed to prevent synthesis of surface antigen would have to be long term.

3 New Approaches for Therapy of HBV Infection

Despite the ability of HBV to evade successfully host defense mechanisms, clearance of chronic hepadnaviral infection does occur spontaneously at a rate of 2.3%–25% a year (VIOLA et al. 1981; HOOFNAGLE et al. 1981; LIAW et al. 1983). Chronic infections that exhibit low viremia and high levels of serum enzymes diagnostic for hepatitis tend to clear spontaneously and are more amenable to antiviral treatment with interferon or inhibitors of DNA replication (SCULLARD et al. 1981; BROOK et al. 1989). As the hepatitis associated with HBV infection at clearance is considered to be caused by the cellular immune response, resolution of infection may be due primarily to changes in the balance of components of the immune system, though a concomitant lessening of ongoing infection within the hepatocytes cannot be discounted. Long-term carriers of HBV frequently make only incomplete virus particles, having ceased to generate replicative DNA intermediates. In the duck hepatitis B virus-(DHBV)-duck model, viral titers in congenital carriers with no liver disease are known to diminish with time. Therefore, in this hepadnaviral system with little or no detectable immune response, viral replication appears to decrease spontaneously with increasing time of infection.

From the foregoing there is reason to believe that one or a combination of approaches to affect viral replication and/or host response could be designed to clear infectious virus from all individuals with active viral replication. However, as eradication may take time, it is important to design therapies that hinder viral replication alone and do not interfere with normal host processes. The currently available treatments with inhibitors of DNA synthesis and with interferons do not yet meet these criteria (DI BISCEGLIE and HOOFNAGLE 1989).

The advances in our understanding of HBV at the molecular level have provided potential strategies to block the multiple steps of HBV replication. In vitro systems are now available for mutational analysis of the small hepadnaviral genomes. This has allowed precise location of functional sites on both nucleic acids and gene products. With knowledge of these important sites, inhibiting or competing molecules can be designed to affect viral and not cellular nucleic acids and proteins. While antivirals which mimic one nucleotide or one amino acid might also affect cellular functions, antivirals made up of multiple nucleotide or amino acid sequences homologous to those in viral nucleic acids and proteins should specifically inhibit steps in viral replication without harming the host.

Oligonucleotides designed to bind to specific regions of RNA or DNA or protein have recently been shown to inhibit replication in other viral families (SMITH et al. 1986; ZERIAL et al. 1987; MATSUKURA et al. 1989; AGRAWAL et al. 1989). They are most effective if made of nucleotide analogues which are resistant to nucleases and of a length to facilitate entry into a cell and movement to the proper intracellular location. Molecular tagging can specifically target the oligonucleotides (LETSINGER et al. 1989), and covalently linked attachments can be used to inactivate the bound nucleic acid (reviewed by STEIN and COHEN 1988).

Peptides homologous to strategic recognition or structural sites have been shown to disrupt normal interactions between proteins or their modifying components (GARTNER and BENNETT 1985; COHEN et al. 1986; reviewed in HASELTINE 1989; MEEK et al. 1990). Inhibiting peptides, if used in multiple doses, should be constructed to avoid the immune response, either by using a sufficiently low number of amino acids to reduce the immunogenic potential or by delivering the peptides to hepatocytes within liver-targeted liposomes, which could contain other drugs as well (reviewed by RANADE 1989).

The most acceptable and convenient mode of delivery of any antiviral is oral, as lengthy treatment may be required for effective eradication of infection. However, administration of other antivirals by injection may be required in addition to a drug taken orally. Delivery of drugs by liposomes or other carrier molecules designed to move primarily to the liver (FIUME et al. 1986; CANONICO et al. 1988) might be used to avoid unacceptable drug side effects in nonhepatic tissues as well as to permit lower concentrations of the drug to be given.

4 The HBV Replication Cycle and Potential Targets for New Antivirals

Some of the steps of the life cycle of the hepadnaviruses summarized in Table 1 are affected by known antiviral compounds. A number of them may be susceptible to the newer types of compounds now available.

Table 1. The life cycle of hepadnaviruses

Free virus
Attachment
Penetration and removal of envelope
Nuclear translocation and nucleocapsid removal
Synthesis and maintenance of CCC DNA
Transcription
Translation
Packaging of viral pregenome
Viral DNA synthesis
Nucleocapsid maturation (nuclear translocation)
Envelopment of nucleocapsids
Export of mature virions

4.1 Free Virus

In infected patients, free or extracellular virus is present in the blood and other body fluids. If the DNA polymerase known to be maintained within the virion proves to be essential to the conversion of incoming relaxed circular DNA to CCA DNA, free virus might be rendered noninfectious by treatment with a drug inhibiting or inactivating this enzyme. Other possibilities for the inactivation of circulating virus might include administering small peptides or other compounds which could compete with and disrupt surface antigen assembly sites. A difficulty in this approach and in any other in which surface antigen components are the antiviral point of attack is that complete virions in the blood are far outnumbered by defective particles containing viral envelope only. These plentiful defective particles will serve as competitors for the therapeutic target and may make this approach less effective.

4.2 Attachment

The binding sites of the surface antigens to hepatocytes are not precisely known. Candidate regions in the preS1 domain which bind to hepatocytes or liver-derived cell lines have been proposed by two groups (NEURATH et al. 1986; PONTISSO et al. 1989). The cellular receptor for HBV has not been identified, nor has it been determined that there is only one receptor. PEEPLES and coworkers (1987) have noted that defective virus particles containing the S domain and not preS bind to and are internalized by cells derived from monkey kidney. Studies of attachment of surface antigen proteins to liver-derived cells are ongoing in several laboratories and should eventually lead to the identification of the cellular receptor and the amino acid regions of the envelope antigens that are involved in binding. As with other virus systems, peptides homologous to either the attachment sites on the virus or on the receptor are candidates for blocking virus attachment, as are any other compounds which might interfere with this step. An example of such a compound may be suramin, which has been found to prevent the initiation of infection of primary duck hepatocytes with DHBV (PETCU et al. 1988). It is not known whether the inhibition is at the level of attachment or at the next stage of the life cycle, penetration and uncoating.

4.3 Penetration and Removal of Envelope

No information is presently available about the entry of the hepadnaviruses into the cell. If the entry mechanism is similar to that of other enveloped viruses, HBV would enter the cytoplasm via clathrin-coated pits and vesicles. Fusion of the envelope with the vesicle membrane following an increased acidity of the vacuole would allow release of the nucleocapsid into the cytoplasm. This process could be arrested by antibody raised to areas on the viral surface antigen essential to uncoating, possibly by the aforementioned suramin, by drugs which change the pH of the intracellular vesicles, such as amantadine and rimantadine, or by other drugs which block or interfere with the uncoating process (for review see DIANA et al. 1989). Drugs affecting this and previous steps of infection initiation would not be expected to eradicate infection alone, but rather prevent reinfection while other drugs hindered subsequent replication steps.

4.4 Nuclear Translocation and Nucleocapsid Removal

At this point either the complete nucleocapsid or the viral DNA alone is moved to the nucleus for the generation or completion of CCC DNA from the incoming virion DNA. It is possible that the nucleocapsids or cores enter the nucleus before the virion DNA is exposed, as core antigen is frequently found in the nucleus in HBV infection, and occasionally with those of the other hepadnaviruses (reviewed in GANEM and VARMUS 1987). If the nucleocapsid moves to the nuclear membrane or into the nucleus, there must be a protein signal for such a translocation (ADAM et al. 1989). This signal would be a prime candidate for antiviral intervention, as preventing movement of nucleocapsids to the nucleus would stop both initiation of infection and the intracellular amplification of infection by return of newly synthesized viral DNA in nucleocapsids to the nucleus. For the same reasons, an equally important therapeutic target would be prevention of nucleocapsid disassembly again by competing peptides, specific antibodies, or other drugs. The blocking effect of interferon on SV40 replication is at this uncoating step (SAMUEL 1988). Such prevention of cellular reinfection might allow the CCC DNA to decay.

4.5 Synthesis and Maintenance of CCC DNA

CCC DNA has been detected only in the nucleus. It is not known whether a viral or a cellular enzyme completes the synthesis of the viral DNA plus strand, although evidence for the use of cellular enzyme is indicated by studies of the effect of phosphonoformate on DHBV infection of hepatocytes (MASON et al. 1987). In the presence of the drug, CCC DNA was synthesized following infection, but no other new viral DNA was made until the drug was removed. Since the viral polymerase is known to be inhibited by phosphonoformate, repair of the incomplete virion DNA could have occurred by a cellular polymerase unaffected by the drug. Subsequent steps of removal of covalently bound protein and the RNA plus-strand primer, ligation, and supercoiling are likely to be done in the nucleus by cellular enzymes, as CCC DNA has not been observed in untreated virions or in those in which DNA polymerization is allowed to proceed in vitro. Prevention of these steps of CCC DNA formation, if they are mediated by cellular enzymes, would be likely to result in unacceptable effects on normal cellular processes. Specific attack on the CCC DNA may be possible, however, through use of competing oligonucleotides that might render the structure less stable during transcription or which could form triple helices susceptible to specific cleavage (reviewed by STEIN and COHEN 1988; FRANCOIS et al. 1989). Any attack on this crucial step of hepadnaviral infection should facilitate eventual viral clearance.

4.6 Transcription

Viral transcription is currently understood to be conducted entirely by cellular elements, which are unsuitable targets for therapy. Though splicing of HBV mRNAs has been observed, it has not been shown to play an essential role in replication and is not a likely point of specific antiviral attack (SU et al. 1989; CHEN et al. 1989). Any compound that would specifically increase the turnover of viral mRNA or prevent the polyA tailing of HBV messages or movement of the pregenome to the nucleocapsid (if this is required) would reduce the level of infection. Another means of lowering the level of replication would be to reduce or shut down the still undefined mechanism for overriding the termination signal at the first readthrough, which allows generation of the greater-than-genome length pregenome and core and precore transcripts.

A viral product known to affect the level of viral transcription in the cell is the transcriptional transactivator encoded by the X-gene of HBV. The product of the X-gene is associated with significantly higher levels of viral replication and consequently may be essential to the maintenance of HBV infection. This *trans*-activator appears to exert its effect by interaction with cellular components rather than binding to viral nucleic acids (TWU and ROBINSON 1989). Therefore, either inhibition of translation of the X-gene ORF by oligonucleotides or inhibition of essential structural or interactive sites of the X-gene product by competing peptides or other drugs might significantly lower intracellular levels of the X-gene product and result in reduction of viral replication with little toxic effect on the body.

4.7 Translation and Posttranslational Modification

The antiviral effect of interferon, which has been shown to reduce HBV titers in patients, is primarily at this replication step in most but not all of the viruses studied (SAMUEL 1988). All of the HBV gene products might be similarly inhibited at the translational level by sense or antisense oligonucleotides. Viral synthesis may be most effectively inhibited by using compounds directed at the least abundant of the viral gene products, such as the *pol* gene products and the X-gene protein. As described in previous chapters, the *pol* gene encodes three activities and functions: reverse transcriptase, RNAseH activity, and the terminal protein thought to serve as primer of minus-strand synthesis. Whether these activities reside in a single molecule or in individual molecules is still unknown, although specific regions of the gene essential for each of these functions are being defined (BARTENSCHLAGER and SCHALLER 1988; RADZIWILL et al. 1990). The stability of the *pol* gene products might be affected by competing peptides or other drugs.

Little is known of posttranslational modification of the X- and *pol* gene products. Recent evidence from J.S. WU and W.S. ROBINSON (personal communication) indicates that the X-gene product is a protein kinase which can self-phosphorylate. This activity, if essential to the function of the X product, is a possible antiviral target. A viral protease has been postulated to be encoded by the C open reading frame (ORF) (MILLER 1987), but no function for this activity has been substantiated (NASSAL et al. 1989; JEAN-JEAN et al. 1989). No viral protein essential to the life cycle of HBV is known to require proteolytic cleavage for activity or maturation. Therefore, unlike the related retroviruses, protease inhibitors do not appear to be important prospects for antihepadnaviral therapy. The more abundant structural proteins, the nucleocapsid and surface antigen proteins, become phosphorylated (nucleocapsid) (ALBIN and ROBINSON 1980; PUGH et al. 1989; SCHLICHT et al. 1989), myristilated (preS1) (PERSING et al. 1987), and glycosylated (preS and S proteins) (reviewed by NEURATH and KENT 1988). Inhibition of any of these modifications might reduce the levels of either circulating virus or infectious virus. A specific inhibitor of core antigen phosphorylation might have a more significant effect upon intracellular viral replication than the surface antigen inhibitors, as inhibition of core phosphorylation could affect the intracellular amplification of CCC DNA as well as the release of mature virus. Although the role of nucleocapsid phosphorylation in the life cycle is not known, mature nucleocapsids in circulating virions are not phosphorylated, suggesting that the phosphorylation may be essential to replication.

4.8 Packaging of Viral Genome

The shortest of the greater-than-genome-length viral mRNAs is encapsidated into assembling core particles (LIEN et al. 1986; SEEGER et al. 1986; WILL et al. 1987; ENDERS et al. 1987). The general localization of the packaging signal on the pregenome is now known (D. GANEM et al., personal communication) and will no doubt be delimited more finely in the near future. The pregenome recognition sites on the nucleocapsid protein are currently under study. Recognition areas for packaging on both the pregenome and nucleocapsid are important possible therapy targets. Blocking packaging would inhibit an important replication step as well as expose the pregenome to degradative enzymes.

4.9 Viral DNA Synthesis

Viral DNA synthesis has been the area of the HBV life cycle targeted by the majority of the drugs tested both in patients and in model systems. All drugs tested which affect this function have some known toxicity. Two treatments tested in patients involved adenine arabinoside or its monophosphate derivative and acyclovir, both of which resulted in transient declines in the titer of HBV viremia. Nucleoside analogues must be converted to nucleoside triphosphates by cellular enzymes to be effective and should not be rapidly catabolized or converted into more toxic compounds. An increasing number of thse analogues and reverse transcriptase inhibitors have been and are being tested in model systems to be described below, though none have proven to eliminate infection. These inhibitors can reduce HBV DNA synthesis by chain termination (acyclovir, dideoxycytidine), by blocking the polymerase binding site for nucleoside triphosphates, the RNA and DNA template binding sites, or the catalytic site for polymerization, and by affecting nucleoside transport and metabolism. Hopefully, testing in model systems and/or better knowledge of the active sites and mechanisms of reverse transcription may allow determination of drugs that will affect HBV reverse transcription alone and have no associated toxicity.

Other functional targets for antiviral therapy in viral DNA synthesis are the terminal protein bound to the 5′ end of the minus DNA strand and thought to serve as primer of its synthesis and the ribonuclease H activity necessary for DNA plus-strand synthesis. Inhibition of DNA synthesis may be possible through competition of binding of the terminal protein to the pregenome either with sense or antisense oligonucleotides or by competing peptides or other drugs. RNaseH inhibitors may be found which block binding of the DNA/RNA substrate or the catalytic site.

4.10 Nucleocapsid Maturation and Envelopment and Virion Export

It is not known whether additional changes in the nucleocapsid structure occur when DNA synthesis is completed and the capsid is ready to be either enveloped and exported or returned to the nucleus to amplify the intranuclear copies of

CCC DNA. It is possible that nucleocapsid modification occurs during envelopment or transport to the nucleus. If there prove to be one or more steps to this maturation, these would be excellent points of antiviral attack.

The recognition sites for envelopment of the nucleocapsids have yet to be determined, but these are also potential targets for therapy. Accumulation of nucleocapsids in the hepatocytes may prove to be toxic, however, as suggested by an unpublished in vitro study using HepG2 cells stably transfected with HBV DNA (P. ROINGEARD et al., personal communication). As with most of the other potential antihepadnaviral drugs, testing in model systems should predict whether the blocking of complete HBV synthesis at any point is itself toxic to hepatocytes. Extracellular release of virions is the final step of the viral life cycle which might be hindered, although if the mechanism of export is the same as for the defective particles, this target may be less desirable because of a possibly greater effective dose requirement or because of the known toxicity associated with cellular accumulation of preS1 proteins.

5 Model Systems for Testing of New Drugs Against HBV

In the past few years, a number of new model systems have become available for testing drugs for efficacy against HBV infection. Prior to discovery of WHV in 1978 and GSHV and DHBV in 1980, potential antihepadnaviral drugs were tested in vitro for their ability to inhibit serum viral particle-associated DNA polymerase activity or in patients for the capacity to reduce serum levels of virus-associated DNA polymerase or of HBsAg. In addition to the current use of the animal models to test antivirals in vivo, there are now a number of systems to test antivirals in cell culture, with either complete or partial life cycles available according to the individual system used. A summary of the various systems is listed in Table 2.

Table 2. Model systems for testing antihepadnaviral drugs

	In vivo	In vitro
Complete life cycle	HBV infection of chimpanzees	HBV infection of human or chimpanzee primary hepatocytes
	WHV infection of woodchucks	WHV or GSHV infection of primary woordchuck hepatocytes
	GSHV infection of ground squirrels or woodchucks	DHBV infection of duck hepatocytes
	DHBV infection of ducks, geese or their eggs	
Partial life cycle	Transgenic mice	HBV-producing, liver-derived cell lines containing HBV DNA
	Nude mice with HBV-producing tumors	HBV-stably integrated HBV DNA
		HBV-transiently transfected with HBV DNA

5.1 In Vivo Models

The first animal model of HBV infection, infection of chimpanzees with HBV, has been little used because of the rarity and expense of chimpanzees. WHV infection of woodchucks and GSHV infection of ground squirrels or woodchucks are models in which the viruses and associated diseases have been relatively well characterized. Yet few in vivo studies of antivirals using these systems have been published (NORDENFELT et al. 1982; SMEE et al. 1985; VENKATESWARAN et al. 1987). These animal models have the advantage of employing the viruses most closely related to HBV and the disadvantage of being relatively inaccessible or expensive for most laboratories. In contrast, the equally well-characterized DHBV-duck model is available around the world and has consequently been that most used to test potential antivirals (reviewed by ZUCKERMAN 1987). Transient reductions in serum DHBV DNA or DNA polymerase levels have been observed following treatment with phosphonoformate (SHERKER et al. 1986), adenine arabinoside (HIROTA et al. 1986, 1987), acyclovir and suramin (TSIQUAYE et al. 1986), dideoxycytidine (KASSIANIDES et al. 1989), and other dideoxynucleosides (LEE et al. 1989). A sustained reduction was observed by TSIQUAYE and colleagues (1986) using suramin injection following acyclovir treatment. Drawbacks to the DHBV-duck system are that DHBV is less like HBV than WHV and GSHV and that avian metabolism and pharmacokinetics of the tested drugs may be unlike those of mammals. An example of the latter is the different in vivo dosages required to affect serum virus titers. In humans or squirrels, doses of 15–20 mg/kg daily of adenine arabinoside are effective, and greater doses are toxic (SACKS et al. 1982; SMEE et al. 1985). In ducks, doses up to 80 mg/kg daily can be given with little toxicity and relatively little effect on serum virus titer (HIROTA et al. 1986; KASSIANIDES et al. 1989; MARION et al., unpublished work). Yet adenine arabinoside is an effective inhibitor of DHBV replication in vitro, causing a 50% reduction in intracellular viral replication at 3 μg/ml (P.L. MARION et al., unpublished observations).

In vivo model systems with complete virus life cycles offer the advantage of being suitable for testing drug delivery methods, pharmacokinetics, drug toxicity, and immune modulators, as well as for testing drugs that affect all stages of the life cycle. Somewhat more limited in vivo models that could be useful for many types of drug tests are transgenic mice which express complete virus (FARZA et al. 1988; ARAKI et al. 1989) or specific viral antigens (CHISARI et al. 1987; BURK et al. 1988) and nude mice with HBV-producing tumors. These systems, however, are not appropriate for testing of the life-cycle steps preceding transcription.

5.2 In Vitro Models

Well-documented hepadnavirus infection of cultured cells which yields infectious progeny is a relatively recent development in the hepadnavirus field. TUTTLEMAN and coworkers (1986b) demonstrated that DHBV infection of primary duck hepatocytes was feasible and reproducible. However, the cultured cells were only susceptible to virus infection for a few days after plating, and multiple cycles of infection were not observed. Use of dimethyl sulfoxide (DMSO) in the culture medium was later found to maintain the differentiated state of the hepatocytes for weeks and permit spread of virus infection to most of the cultured cells (PUGH and SUMMERS 1989; GALLE et al. 1989). For those with access to DHBV-infected flocks, it is possible to culture virus-producing primary hepatocytes taken from congenitally infected ducklings and avoid problems of low infectivity of inoculum and the relatively restricted initial infection seen even with high titer inocula.

Recently, successful mammalian hepadnavirus infection of cultured hepatocytes has become feasible, if less available to most laboratories than cell culture of DHBV. ALDRICH and colleagues (1989) were able to grow both WHV and GSHV in primary hepatocyte cultures from woodchucks. HBV has been shown to enter and replicate in primary cultures of both fetal and adult hepatocytes (OCHIYA et al. 1989; RIJNTJES et al. 1988; GRIPON et al. 1988) and appears to infect the hepatoblastoma cell line HepG2 to a more limited extent (M. Petit, personal communication).

The in vitro DHBV-duck system has already been useful to study the efficacy and mechanism of action on DHBV replication of a number of drugs (MASON et al. 1987; PETCU et al. 1988; LEE et al. 1989). All of these culture systems, but especially that of the more available duck model, should be used with increasing frequency for screening of potential drugs affecting any phase of the hepadnaviral life cycle.

For in vitro study of viral replication steps from transcription through virus release, it is now possible to use cell lines derived from well-differentiated hepatomas (HUH-7 and others) and hepatoblastomas (HepG2, Huh-6), in which infectious virus can be grown following transient or stable transfection with HBV (or other hepadnaviral) DNA from which the complete pregenome can be transcribed (SUREAU et al. 1986; SELLS et al. 1987; YAGINUMA et al. 1987). Cell lines with stably integrated HBV DNA and which produce infectious virus are particularly convenient for testing compounds which might affect viral reverse transcription and plus-strand DNA synthesis (UEDA et al. 1989; PRICE et al. 1989).

6 Conclusions and Outlook

Despite a number of characteristics of the HBV life cycle which contribute to its tendency toward persistent infection and resistance to drugs affecting nucleic acid synthesis, the outlook for developing therapy for chronic HBV infection has improved considerably in the past several years. The key to this improvement is the explosion of information concerning the functionally important nucleic acid and protein sites in the virus life cycle due to recently discovered in vitro systems with which to test mutations of the viral genome. This new knowledge, combined with the increasing availability of new types of antiviral compounds designed to compete with or inhibit specific viral sequences and the newer accessible in vivo and in vitro systems with which to test the drugs, provides encouraging new prospects for the therapy of HBV in the immediate future. Finally, other, more novel possibilities for therapy can be envisaged. These include development of dominant negative HBV mutants which do not produce mature virions but defective gene products which interfere with wild-type virus replication (HERSKOWITZ 1987) and use of ribozymes to cleave HBV RNA or DNA (HERSCHLAG and CECH 1990). Such ribozymes may be developed from hepatitis delta virus, which infects human hepatocytes and already possesses RNA self-cleavage capabilities.

References

Adam SA, Lobl TJ, Mitchell MA, Gerace L (1989) Identification of specific binding proteins for a nuclear location sequence. Nature 337: 276–279

Agarwal S, Ikeuchi T, Sun D, Sarin PS, Konopka A, Maizel J, Zamecnik PC (1989) Inhibition of human immunodeficiency virus in early infected and chronically infected cells by antisense oligodeoxynucleotides and their phosphorothioate analogues. Proc Natl Acad Sci USA 86: 7790–7794

Albin C, Robinson WS (1980) Protein kinase activity in hepatitis B virus. J Virol 34: 297–302

Aldrich CE, Coates L, Wu TT, Newbold J, Tennant BC, Summers J, Seeger C, Mason WS (1989) *In vitro* infection of woodchuck hepatocytes with woodchuck hepatitis virus and ground squirrel hepatitis virus. Virology 172: 247–252

Alexander GJ, Williams R (1988) Natural history and therapy of chronic hepatitis B virus infection. Am J Med 85: 143–146

Araki K, Miyazaki J, Hino O, Tomita N, Chisaka O, Matsubara K, Yamamura K (1989) Expression and replication of hepatitis B virus genome in transgenic mice. Proc Natl Acad Sci USA 86: 207–211

Bartenschlager R, Schaller H (1988) The amino-terminal domain of the hepadnaviral P-gene encodes the terminal protein (genome-linked protein) believed to prime reverse transcription. EMBO J 7: 4185–4192

Beasley RP, Hwang LY (1984) Epidemiology of hepatocellular carcinoma. In: Vyas GN, Dienstag JL, Hoofnagle JH (eds) Viral hepatitis and liver disease. Grune and Stratton, New York, pp 209–224

Brook MG, Karayiannis P, Thomas HC (1989) Which patients with chronic hepatitis B virus infection will respond to alpha-interferon therapy? A statistical analysis of predictive factors. Hepatology 10: 761–763

Burk RD, DeLoia JA, elAwady MK, Gearhart JD (1988) Tissue preferential expression of the hepatitis B virus (HBV) surface antigen gene in two lines of HBV transgenic mice. J Virol 62: 649–654

Canonico PG, Kende M, Gabrielsen B (1988) Carrier-mediated delivery of antiviral agents. Adv Virus Res 35: 271–312

Chen PJ, Chen CR, Sung JL, Chen DS (1989) Identification of a doubly spliced viral transcript joining the separated domains for putative protease and reverse transcriptase of hepatitis B virus. J Virol 63: 4165–4171

Chisari FV, Filippi P, Buras J, McLachlan A, Popper H, Pinkert CA, Palmiter RD, Brinster RL (1987) Structural and pathological effects of synthesis of hepatitis B virus large envelope polypeptide in transgenic mice. Proc Natl Acad Sci USA 84: 6909–6913

Cohen EA, Gaudreau P, Brazeau P, Langelier Y (1986) Specific inhibition of herpesvirus ribonucleotide reductase by a nonapeptide derived from the carboxy terminus of subunit 2. Nature 321: 441–443

Diana GD, Pevear DC, Otto MJ, McKinlay MA, Rossmann MG, Smith T, Badger J (1989) Inhibitors of viral uncoating. Pharmacol Ther 42: 289–305

Di Bisceglie AM, Hoofnagle JH (1989) Treatment of chronic type B hepatitis, Cancer Detection and Prevention 14: 291–293

Dunsford HA, Sell S, Chisari FV (1990) Hepatocarcinogenesis due to chronic liver cell injury in hepatitis B virus transgenic mice. Cancer Research 50: 3400–3407

Enders GH, Ganem D, Varmus HE (1987) 5′-terminal sequences influence the segregation of ground squirrel hepatitis virus RNAs into polyribosomes and viral core particles. J Virol 61: 35–41

Farza H, Hadchouel M, Scotto J, Tiollais P, Babinet C, Pourcel C (1988) Replication and gene expression of hepatitis B virus in a transgenic mouse that contains the complete viral genome. J Virol 62: 4144–4152

Fiume L, Bassi B, Busi C, Mattioli A, Spinosa G (1986) Drug targeting in antiviral chemotherapy. Biochem Pharmacol 35: 967–972

Francois JC, Saison-Behmoaras T, Barbier C, Chassignol M, Thuong NT, Helene C (1989) Sequence-specific recognition and cleavage of duplex DNA via triple-helix formation by oligonucleotides covalently linked to a phenanthroline-copper chelate. Proc Natl Acad Sci USA 86: 9702–9706

Galle PR, Schlicht HJ, Kuhn C, Schaller H (1989) Replication of duck hepatitis B virus in primary duck hepatocytes and its dependence on the state of differentiation of the host cell. Hepatology 10: 459–465

Ganem D, Varmus HE (1987) The molecular basis of the hepatitis B viruses. Annu Rev Biochem 56: 651–693

Garcia G, Gentry KR (1989) Chronic viral hepatitis. Med Clin North Am 73: 971–983

Gartner TK, Bennett JS (1985) The tetrapeptide analogue of the cell attachment site of fibronectin inhibits platelet aggregation and fibrinogen binding to activated platelets. J Biol Chem 260: 11891–11894

Gripon P, Diot C, Theze N, Fourel I, Loreal O, Brechot C, Guguen-Guillouzo C (1988) Hepatitis B virus infection of adult human hepatocytes cultured in the presence of dimethyl sulfoxide. J Virol 62: 4136–4143

Haseltine WA (1989) Development of antiviral drugs for the treatment of AIDS: strategies and prospects. J Acquired Immune Deficiency Syndromes 2: 311–334

Herschlag D, Cech, TR (1990) DNA cleavage catalysed by the ribozyme from *Tetrahymena*. Nature 344: 405–409

Herskowitz I (1987) Functional inactivation of genes by dominant negative mutations. Nature 329: 219–222

Hirota K, Sherker AH, Omata M, Yokosuka O, Okuda K (1986) Effects of adenine arabinoside and corticosteroid on replication of duck hepatitis B virus DNA in the liver. Ann Acad Med Singapore 15: 227–232

Hirota K, Sherker AH, Omata M, Yokosuka O, Okuda K (1987) Effects of adenine arabinoside on serum and intrahepatic replicative forms of duck hepatitis B virus in chronic infection. Hepatology 7: 24–28

Hoofnagle JH, Dusheiko GM, Seeff LB, Jones EA, Waggoner JG, Buskell Bales Z (1981) Seroconversion from hepatitis B virus e antigen to antibody in chronic type B hepatitis. Ann Intern Med 94: 744–748

Jean-Jean O, Salhi S, Carlier D, Elie C, De Recondo AM, Rossignol JM (1989) Biosynthesis of hepatitis B virus e antigen: directed mutagenesis of the putative aspartyl protease site. J Virol 63: 5497–5500

Kassianides C, Hoofnagle JH, Miller RH, Doo E, Ford H, Broder S, Mitsuya H (1989) Inhibition of duck hepatitis B virus replication by 2′,3′-dideoxycytidine. A potent inhibitor of reverse transcriptase. Gastroenterology 97: 1275–1280

Korba BE, Cote PJ, Shapiro M, Purcell RH, Gerin JL (1989) *In vitro* production of infectious woodchuck hepatitis virus by lipopolysaccaride-stimulated peripheral blood lymphocytes. J Infect Dis 160: 572–576

Lee B, Luo WX, Suzuki S, Robins MJ, Tyrrell DL (1989) *In vitro* and *in vivo* comparison of the abilities of purine and pyrimidine 2′,3′-dideoxynucleosides to inhibit duck hepadnavirus. Antimicrob Agents Chemother 33: 336–339

Letsinger RL, Zhang GR, Sun DK, Ikeuchi T, Sarin PS (1989) Cholesteryl-conjugated oligonucleotides: synthesis, properties, and activity as inhibitors of replication of human immunodeficiency virus in cell culture. Proc Natl Acad Sci USA 86: 6553–6556

Lever AM (1988) Treatment of the chronic hepatitis B virus carrier state. J Infect 16: 221–229

Liaw YF, Chu CM, Su IJ, Huang MJ, Lin DY, Chang-Chien CS (1983) Clinical and histological events preceding hepatitis B e antigen seroconversion in chronic type B hepatitis. Gastroenterology 84: 216–219

Lien JM, Aldrich CE, Mason WS (1986) Evidence that a capped oligoribonucleotide is the primer for duck hepatitis B virus plus-strand DNA synthesis. J Virol 57: 229–236
Marion PL (1988) Use of animal models to study hepatitis B virus. Prog Med Virol 35: 43–75
Mason WS, Lien J-M, Petcu DJ, Coates L, London WT, O'Connell A, Aldrich C, Custer RP (1987) *in vivo* and *in vitro* studies on duck hepatitis B virus replication. In: Hepadna viruses. UCLA symposia on molecular and cellular biology, new series. Liss, New York
Matsukura M, Zon G, Shinozuka K, Robert-Guroff M, Shimada T, Stein CA, Mitsuya H, Wong-Staal F, Cohen JS, Broder S (1989) Regulation of viral expression of human immunodeficiency virus *in vitro* by an antisense phosphorothioate oligodeoxynucleotide against rev (art/trs) in chronically infected cells. Proc Natl Acad Sci USA 86: 4244–4248
Maynard JE, Kane MA, Hadler SC (1989) Global control of hepatitis B through vaccination: role of hepatitis B vaccine in the Expanded Programme on Immunization. Rev Infect Dis 11: S574–S578
Meek TD, Lambert DM, Dreyer GB, Carr TJ, Tomaszek TA Jr, Moore ML, Strickler JE, Debouck C, Hyland LJ, Matthews TJ, Metcalf BW, Petteway SR (1990) Inhibition of HIV-1 protease in infected T-lymphocytes by synthetic peptide analogues. Nature 343: 90–92
Miller RH (1987) Proteolytic self-cleavage of hepatitis B virus core protein may generate serum e antigen. Science 236: 722–725
Nagaya T, Nakamura T, Tokino T, Tsurimoto T, Imai M, Mayumi T, Kamino K, Yamamura K, Matsubara K (1987) The mode of hepatitis B virus DNA integration in chromosomes of human hepatocellular carcinoma. Genes Dev 1: 773–782
Nassal M, Galle PR, Schaller H (1989) Proteaselike sequence in hepatitis B virus core antigen is not required for e antigen generation and may not be part of an aspartic acid-type protease. J Virol 63: 2598–2604
Neurath AR, Kent SBH (1988) The pre-S region of hepadnavirus envelope proteins. Adv Virus Res 34: 65–142
Nordenfelt E, Widell A, Hansson BG, Lofgren B, Moller-Nielsen C, Oberg B (1982) No *in vivo* effect of trisodium phosphonoformate on woodchuck hepatitis virus production. Acta Pathol Microbiol Immunol Scand 90: 449–451
Ochiya T, Tsurimoto T, Ueda K, Okubo K, Shiozawa M, Matsubara K (1989) An *in vitro* system for infection with hepatitis B virus that uses primary human fetal hepatocytes. Proc Natl Acad Sci USA 86: 1875–1879
Peeples ME, Komai K, Radek R, Bankowski MJ (1987) A cultured cell receptor for the small S protein of hepatitis B virus. Virology 160: 135–142
Persing DH, Varmus HE, Ganem D (1987) The preS1 protein of hepatitis B virus is acylated at its amino terminus with myristic acid. J Virol 61: 1672–1677
Petcu DJ, Aldrich CE, Coates L, Taylor JM, Mason WS (1988) Suramin inhibits *in vitro* infection by duck hepatitis B virus, Rous sarcoma virus, and hepatitis delta virus. Virology 167: 385–392
Pontisso P, Petit M-A, Bankowski MJ, Peeples ME (1989) Human liver plasma membranes contain receptors for the hepatitis B virus pre-S1 region and, via polymerized human serum albumin for the pre-S2 region. J Virol 63: 1981–1988
Price PM, Banerjee R, Acs B (1989) Inhibition of the replication of hepatitis B virus by the carbocyclic analogue of 2′-deoxyguanosine. Proc Natl Acad Sci USA 86: 8541–8544
Pugh JC, Summers JW (1989) Infection and uptake of duck hepatitis B virus by duck hepatocytes maintained in the presence of dimethyl sulfoxide. Virology 172: 564–572
Pugh J, Zweidler A, Summers J (1989) Characterization of the major duck hepatitis B virus core particle protein. J Virol 63: 1371–1376
Radziwill G, Tucker W, Schaller H (1990) Mutational analysis of the hepatitis B virus P gene product: domain structure and RNase H activity. J Virol 64: 613–620
Ranade VV (1989) Drug delivery systems. 1. Site-specific drug delivery using liposomes as carriers. J Clin Pharmacol 29: 685–694
Rijntjes PJ, Moshage HJ, Yap SH (1988) *In vitro* infection of primary cultures of cryopreserved adult human hepatocytes with hepatitis B virus. Virus Res 10: 95–109
Sacks SL, Scullard GH, Pollard RB et al. (1982) Antiviral treatment of chronic hepatitis B virus infection: pharmacokinetics and side effects of interferon and adenine arabinoside alone and in combination. Antimicrob Agents Chemother 21: 93–100
Samuel CE (1988) Mechanisms of the antiviral action of interferons. Prog Nucleic Acid Res Mol Biol 35: 27–72
Schlicht HJ, Bartenschlager R, Schaller H (1989) The duck hepatitis B virus core protein contains a

highly phosphorylated C terminus that is essential for replication but not for RNA packaging. J Virol 63: 2995–3000
Scullard GH, Pollard RB, Smith JL, Sacks SL, Gregory PB, Robinson WS, Merigan TC (1981) Antiviral treatment of chronic hepatitis B virus infection. I. Changes in viral markers with interferon combined with adenine arabinoside. J Infect Dis 143: 772–783
Seeger C, Ganem D, Varmus HE (1986) Biochemical and genetic evidence for the hepatitis B virus replication strategy. Science 232: 477–484
Sells MA, Chen ML, Acs G (1987) Production of hepatitis B virus particles in HepG2 cells transfected with cloned hepatitis B virus DNA. Proc Natl Acad Sci USA 84: 1005–1009
Sherker AH, Hirota K, Omata M, Okuda K (1986) Foscarnet decreases serum and liver duck hepatitis B virus DNA in chronically infected ducks. Gastroenterology 91: 818–824
Smee DF, Knight SS, Duke AE, Robinson WS, Matthews TR, Marion PL (1985) Activities of arabinosyladenine monophosphate and 9-(1,3-dihydroxy-2-propoxymethyl) guanine against ground squirrel hepatitis virus *in vivo* as determined by reduction in serum virion-associated DNA polymerase. Antimicrob Agents Chemother 27: 277–279
Smith CC, Aurelian L, Reddy MP, Miller PS, Ts'o PO (1986) Antiviral effect of an oligo (nucleoside methylphosphonate) complementary to the splice junction of herpes simplex virus type 1 immediate early pre-mRNAs 4 and 5. Proc Natl Acad Sci USA 83: 2787–2791
Stein CA, Cohen JS (1988) Oligodeoxynucleotides as inhibitors of gene expression: a review. Cancer Res 48: 2659–2668
Su TS, Lai CJ, Huang JL, Lin LH, Yauk YK, Chang CM, Lo SJ, Han SH (1989) Hepatitis B virus transcript produced by RNA splicing. J Virol 63: 4011–4018
Sureau C, Romet-Lemonne J-L, Mullins JI, Essex M (1986) Production of hepatitis B virus by a differentiated human hepatoma cell line after transfection with cloned circular HBV DNA. Cell 47: 37–47
Tsiquaye KN, Collins P, Zuckerman AJ (1986) Antiviral activity of the polybasic anion, suramin and acyclovir in Hepadna virus infection. J Antimicrob Chemother 18: 223–228
Tuttleman JS, Pourcel C, Summers J (1986a) Formation of the pool of covalently closed circular viral DNA in hepadnavirus-infected cells. Cell 47: 451–460
Tuttleman JS, Pugh JC, Summers JW (1986b) *In vitro* experimental infection of primary duck hepatocyte cultures with duck hepatitis B virus. J Virol 58: 17–25
Twu JS, Robinson WS (1989) Hepatitis B virus X gene can transactivate heterologous viral sequences. Proc Natl Acad Sci USA 86: 2046–2050
Ueda K, Tsurimoto T, Nagahata T, Chisaka O, Matsubara K (1989) An *in vitro* system for screening anti-hepatitis B virus drugs. Virology 169: 213–216
Venkateswaran PS, Millman I, Blumberg BS (1987) Effects of an extract from Phyllanthus niruri on hepatitis B and woodchuck hepatitis viruses: *in vitro* and *in vivo* studies. Proc Natl Acad Sci USA 84: 274–278
Viola LA, Barrison IA, Coleman JC, Paradinas FJ, Murray-Lyon IM (1981) The HBe antigen-antibody system and its relationship to clinical and laboratory findings in 100 chronic HBsAg carriers in Great Britain. J. Med Virol 8: 169–175
Will H, Reiser W, Weimer T, Pfaff E, Buscher M, Sprengel R, Cattaneo R, Schaller H (1987) Replication strategy of human hepatitis B virus. J Virol 61: 904–911
Wu T-T, Coates L, Aldrich CE, Summers J, Mason WS (1990) In hepatocytes infected with duck hepatitis B virus, the template for viral RNA synthesis is amplified by an intracellular pathway. Virology 175 (in press)
Yaginuma K, Shirakata Y, Kobayashi M, Koike K (1987) Hepatitis B virus (HBV) particles are produced in a cell culture system by transient expression of transfected HBV DNA. Proc Natl Acad Sci USA 84: 2678–2682
Zerial A, Thuong NT, Helene C (1987) Selective inhibition of the cytopathic effect of type A influenza viruses by oligodeoxynucleotides covalently linked to an intercalating agent. Nucleic Acids Res 15: 9909–9919
Zuckerman AJ (1987) Screening of antiviral drugs for hepadna virus infection in Pekin ducks: a review. J Virol Methods 17: 119–126

Inherited Hepatic Enzyme Defects as Candidates for Liver-Directed Gene Therapy

A. L. HORWICH

There are a small number of inherited hepatic defects that are so devastating that virtually 100% of the affected infants either die or suffer severe developmental disability. For these defects, serious consideration must be given to any new treatment that offers the possibility of altering the extremely poor prognosis. In the desparate clinical setting of these single gene defects, two forms of therapy have recently been considered—hepatic transplantation and gene therapy. Here, I discuss two severe single gene disorders that might be candidates for such novel therapies, outline their natural histories, elaborate on the diagnostic tools available, and describe the result of hepatic transplantation in one case. I conclude that there remains a place for the development of cell or gene therapy for these severe disorders, speculate on potential strategies, and suggest that, if a safe and effective strategy could be developed, other less severe conditions of the liver could ultimately be considered for such therapy.

Howard Hughes Medical Institute and Department of Human Genetics, Yale University School of Medicine, New Haven, CT 06510, USA

Current Topics in Microbiology and Immunology, Vol. 168

1 Ornithine Transcarbamylase Deficiency

1.1 Case History

Our first case concerns a 4200 g full-term male infant, the first baby of a 20-year-old married couple. He was entirely normal, feeding well, rooming in with his mother. Early on the 2nd day of life he began to spit up his feedings. Later on, his respiratory rate was noted to be increased. A chest film was normal, and a blood gas showed a respiratory alkalosis. Shortly after, he became lethargic, then unresponsive, and began to have seizures. Blood ammonium was measured and found to be ten times the upper limit of normal. Despite heroic medical measures his ammonium could not be lowered, he never regained consciousness, and he died on the 4th day of life.

1.2 Diagnosis, Biochemistry, Genetics

Severe hyperammonemia in this newborn infant suggested deficiency of a step of the urea cycle, a metabolic pathway that converts ammonium, a toxic intermediate of amino acid metabolism, to the excretable molecule urea (see Fig. 1 and for review BRUSILOW and HORWICH 1989). At the time hyperammonemia was detected, this infant underwent plasma and urine studies which suggested deficiency of the second step of the cycle, catalyzed by the enzyme ornithine transcarbamylase (OTC). When a postmortem liver specimen was assayed, no OTC activity was detected, whereas activities of the other enzymes of the cycle were normal.

OTC is a trimer of identical 36K subunits that, like the first enzyme of the urea cycle pathway, carbamyl phosphate synthetase I (CPS I), is expressed specifically in the liver and enterocytes, and resides in mitochondria. Parenthetically, CPS I deficiency presents a clinical picture virtually identical to that of OTC deficiency. The remaining enzymes of the urea cycle are "constitutively" expressed, in all tissues, and are cytosolic. Inherited defects of these enzymes can also present with hyperammonemia, but in general these disorders respond well to medical management measures discussed below (BRUSILOW and HORWICH 1989). For all of these defects, during intrauterine life, accumulated ammonium can transit the placenta and enter the maternal urea cycle. However, following birth, with detachment from the maternal circulation and with institution of protein feeds, ammonium can rapidly accumulate.

OTC deficiency is inherited as an X-linked trait, consistent with localization of the gene to the short arm of the X chromosome (LINDGREN et al. 1984). As a lethal X-linked condition, Haldane's law would suggest that one-third of cases represent new mutations (HALDANE 1935). Surprisingly, the recent work of Brusilow has shown this not to be the case. Brusilow devised a test that can identify female heterozygote carriers. It identifies carriers by their accumulation

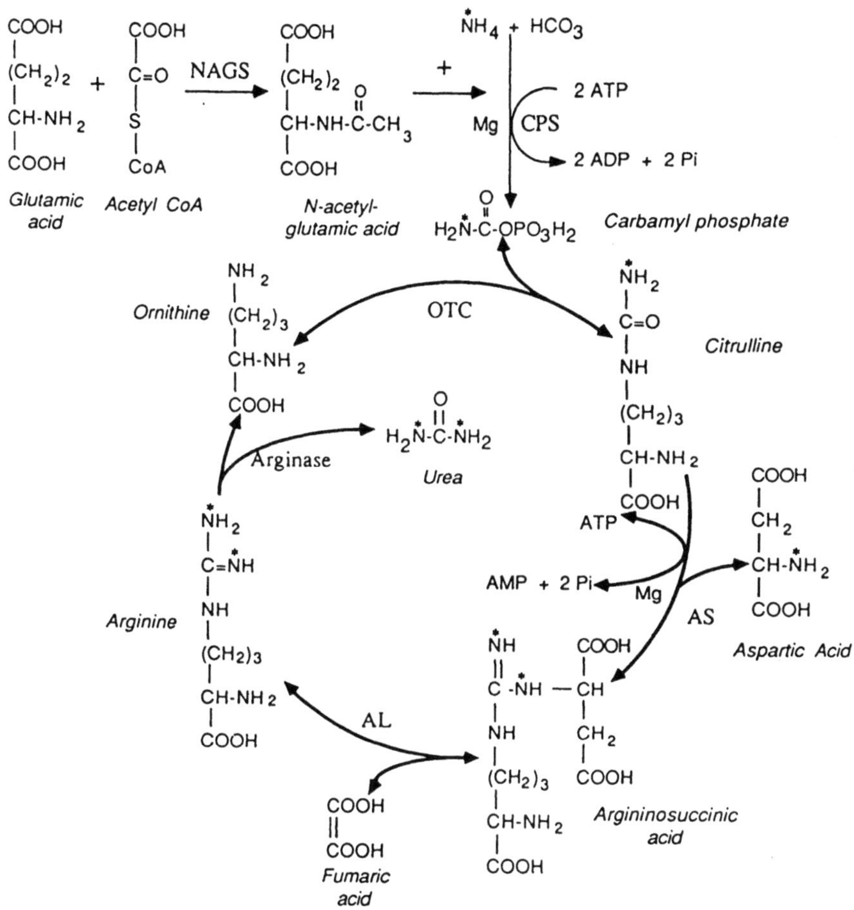

Fig. 1. Urea cycle pathway. *NAGS*, *OTC*, ornithine transcarbamylase; *CPS*, carbamyl phosphate synthetase; *AS*, arginosuccinate synthetase; *AL*, arginosuccinate lyase. (From Brusilow and Horwich 1989, with permission)

of carbamyl phosphate behind the OTC block, and by its subsequent conversion in the cytosol (via the pyrimidine biosynthetic pathway) to orotidine (Houser et al. 1990). When this assay, the so-called allopurinol test, was applied to mothers of first-affected OTC-deficient males, it became clear that most affected males descend from female carrier mothers and that, in fact, mutations at the OTC locus almost always arise in sperm. Where a first affected male infant is seen, the molecular defect is likely to have arisen in the maternal grandfather's sperm. Thus for most women who have had a previous OTC-deficient son, there is a 50% risk for each male pregnancy of being similarly affected with OTC deficiency, and a 50% risk for each female pregnancy of the baby being an OTC heterozygote carrier.

The X-linked pattern of inheritance can lead to multiple neonatal male deaths, but also to occasional symptomatic carrier females. These women have undergone lyonization, the process of random X chromosome inactivation occurring early in female development (LYON 1961), in an unfortunate manner—the majority of the active X chromosomes of the liver bear the mutant OTC gene, whereas the minority bear the wild-type allele. The critical activity level appears to be approximately 15% of normal, based on studies of partially deficient males. When total activity falls below this level, clinical symptomatology is observable. This takes the form of episodic ammonia intoxication, with signs of vomiting and loss of consciousness. The episodes are usually triggered by high dietary protein intake, by intercurrent illness with its attendant catabolic state, and, curiously, by the postpartum state (ARNE et al 1990). Even with residual enzyme activity, these women can die from such episodes, or be left with neurologic injury. In general, however, their mortality and morbidity are much less than that of hemizygous affected males who are devoid of detectable enzyme activity. Symptomatic OTC heterozygous females have been significantly helped by the use of medical treatment for removing ammonia by non-urea cycle-dependent mechanisms.

1.3 Medical Management, Prognosis

Medical treatment of carrier females, and of hemizygous males, exploits latent amino acid acylation pathways. Administration of two agents, sodium benzoate and sodium phenylbutyrate, results in removal of the amino acids glycine and glutamine, respectively, as excretable conjugates, hippurate and phenylacetylglutamine, respectively (BRUSILOW et al. 1980; BATSHAW et al. 1982). Thus these two agents can remove nitrogen by paths that do not involve the urea cycle. The agents can be given long-term in an oral form to symptomatic female heterozygotes, producing improvement of their protein tolerance and prevention of hyperammonemic episodes. They are also of benefit to hemizygous males, although it is unclear whether they significantly alter overall survival rates. In addition to chronic administration as a preventive measure, the two conjugating agents can also be given parenterally during acute episodes to rapidly remove ammonia (BRUSILOW et al. 1984). However, for severe acute hyperammonemic episodes, hemodialysis appears to be the most efficient means of rapid removal of ammonia (SPENCE et al. 1989).

Medical intervention has been quite successful in managing symptomatic OTC female heterozygotes, but it has also been successful in managing patients with CPS I deficiency, including a 2-year-old in New Haven. Here, therapy has involved two major elements: dietary protein restriction, to a level of intake just sufficient to permit growth, without an excess that might overload the residual urea cycle activity, and daily administration, orally, of conjugating agents. While our patient's affected sibling had died on the 3rd day of life from ammonia intoxication, with a clinical course resembling that of the foregoing OTC-deficient case, medical management instituted from the day of birth has been successful in our patient. During pregnancy, her parents had declined molecular genetic prenatal diagnosis to address the 25% chance that she would be

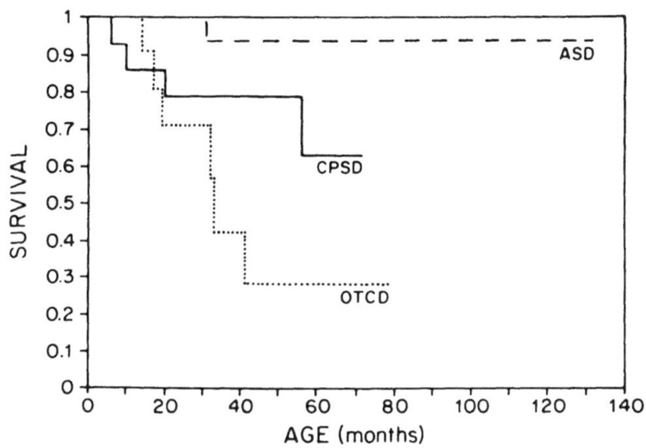

Fig. 2. Survival curve for patients with ornithine transcarbamylase deficiency (*OTCD*), carbamyl phosphate synthetase I deficiency (*CPSD*), and arginosuccinate synthetase deficiency (*ASD*). (From Brusilow and Horwich 1989, with permission)

affected by this autosomal recessive condition. However, the diagnosis was made immediately after birth by measurement of plasma amino acids. Treatment with protein restriction and oral conjugating agents was instituted from that point. She has had two episodes of hyperammonemia requiring emergency treatment with intravenous administration of conjugating agents and has required the insertion of a feeding tube into the stomach in order to maintain her nutritional status in the face of unexplained anorexia. She has otherwise grown and developed normally.

In contrast to this result with CPS I deficiency, OTC deficiency appears to be a much more brittle disorder, that is much less responsive to medical management. This is reflected in the survival curve shown in Fig. 2. While 85% of CPS-deficient patients are alive at age 3.5 year, only 30% of OTC-deficient males survive to this age.

1.4 Molecular Pathology in Man and Mouse

With the molecular cloning of both human OTC cDNA (1400 bp) (HORWICH et al. 1984) and the gene (73 kb divided into 10 exons) (HATA et al. 1988), it has become possible to analyze the defects at the molecular level. The molecular defects that have been characterized at the OTC locus prove to be heterogeneous. Approximately 10% of affected males exhibit gene deletion (ROZEN et al. 1985; FOX and ROSENBERG 1988); the remainder of the defects appear to be point mutations. Some of these mutations affect RNA processing, leading to exon deletion (R. Carstens, to be published) or to unstable RNA R. Carstens, W. Fenton, and L.E. Rosenberg, in press; GROMPE et al. 1989). Other point mutations are associated with effects at the protein level, producing premature chain termination or amino acid substitution (MADDALENA et al. 1988; HATA et al. 1989; GROMPE et al. 1989).

Two strains of mice with inherited OTC deficiency have also been identified. Animals of both strains exhibit a deficiency of fur (sparse fur) in the newborn period, related to the abnormality of protein metabolism, and both exhibit orotic aciduria. However, hemizygous male mice with these defects are viable. This has facilitated molecular characterization of these defects. It has also enabled mating of the affected animals with heterozygous female carrier mice to produce female homozygotes. These animals, when mated in turn with hemizygous males, produce litters that are 100% affected, providing an excellent setting in which to test new therapeutic strategies. The specific defects in the two strains have been characterized. In the mutant allele, called sparse fur (*spf*), an amino acid substitution altering *his* 85 of the mature OTC subunit to asparagine results in production of an enzyme with 25% normal activity at physiologic pH (VERES et al. 1987).

The *spf ash* allele (ash for abnormal skin and hair) carries a base substitution affecting amino acid 97 of the mature OTC subunit, altering it from *arg* to *his* (HODGES and ROSENBERG 1989). The codon for this amino acid is also part of a splice donor site, and a significant percentage of the RNA in these animals fails to splice at this site, using instead a site within the following intron. As a result two products are encoded. One is the product of the spliced RNA, which is present at a level about 5% of wild-type, and contains the residue substitution at position 97. The other contains an insertion of 16 amino acids in frame between the usual positions 97 and 98. The latter product fails to assemble into trimeric enzyme. The former product *can* assemble and produces an enzyme with a normal specific activity. Thus the overall level of activity in *spf ash* mice is about 5% of normal, corresponding to the reduced level of RNA encoding a normally size subunit.

1.5 Molecular Diagnosis

Where a precise molecular defect has been identified at the DNA level in an affected male, it is feasible to directly diagnose carrier status of at-risk females in a pedigree. Also, prenatal diagnosis can be easily carried out on at-risk pregnancies. However, even where the precise defect cannot be identified, which is generally the situation, it is possible in most pedigrees to "track" the defective OTC allele using a number of restriction fragment length polymorphisms (RFLPs) detected at the OTC locus itself (FOX et al. 1986; SPENCE et al. 1989). Thus, where either the specific molecular defect of an affected male can be identified, or where the affected X chromosome observed in an affected male can be distinguished from the normal chromosome upon RFLP analysis of a carrier mother, there is the possibility of prenatal diagnosis for future pregnancies. Such diagnosis can be carried out using chorionic villus tissue, obtained at 7 weeks into pregnancy, or using amniotic fluid cells obtained at 16–18 weeks. Where neither approach of DNA diagnosis could be informative, an additional diagnostic procedure can be pursued, involving biopsy of the fetal liver at 19 weeks pregnancy under ultrasound guidance. Here, a small piece of fetal liver tissue is assayed for OTC enzyme activity (HOLZGREVE and GOLBUS 1984). The drawback to this procedure is the risk of fetal loss from the procedure itself, estimated at 5%.

2 Hereditary Tryrosinemia

2.1 Case Report

The patient was a 3850 g female infant who was well until 3 weeks of age when she stopped taking feeds and developed abdominal distention. On examination she had both umbilical and inguinal hernias and the presence of abdominal fluid, so-called ascites. Laboratory studies revealed her to be in liver failure, and the presence of an elevated plasma tyrosine and of urinary organic compounds parahydroxyphenylpyruvic acid and succinylacetone indicated the diagnosis of hereditary tyrosinemia, which results from deficiency of the hepatic enzyme involved in tyrosine degradation, fumarylacetoacetate hydrolase (FAH). Attempts to treat the baby's condition by dietary restriction of tyrosine, and by medical support of liver function with vitamin K, plasma, platelets, and diuretics, were unsuccessful. The infant developed significant clinical bleeding from her nose and rectum, due to deficiency of clotting factors (resulting from hepatic failure), failed to maintain weight, and became emaciated and lethargic. With a deteriorating clinical picture, at 14 weeks of age, a liver transplant was carried out using the liver of a 6-week-old donor. Within hours of transplantation, the biochemical derangements were no longer detectable. The patient had an uncomplicated postoperative course with the exception of a mild episode of rejection treated with an increased dose of methylprednisolone. She is now 3 years old and has grown and developed normally. She is maintained on three immunosuppressive agents, cyclosporin, prednisone, and immuran.

2.2 Diagnosis, Medical Treatment, Prognosis

The deficiency of hepatic FAH in hereditary tyrosinemia (see Fig. 3 and for review GOLDSMITH and LABERGE 1989) results in the characteristic accumulation of organic acid and amino acid metabolites behind the metabolic obstruction. The enzyme is normally expressed in hepatocytes and in renal tubular cells, but the former comprises the major source of enzymatic activity. The metabolites accreted in the face of deficiency are normally present in vanishing amounts because they are intermediates of side pathways that are not normally employed. The functions of such pathways are unknown. However, in the situation of deficiency of a main pathway enzyme, these pathways become

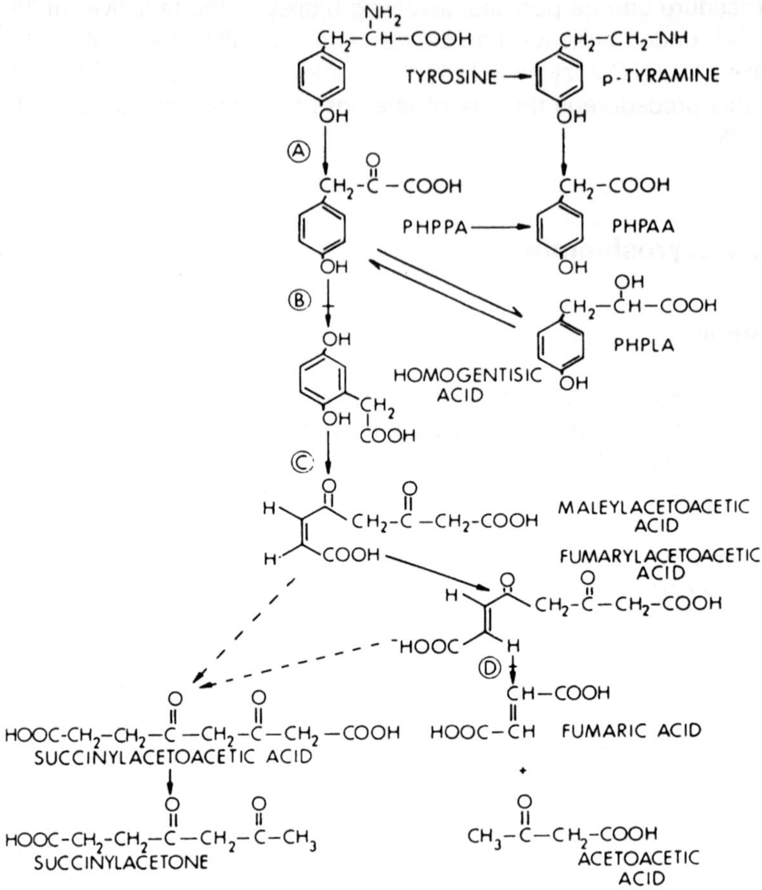

Fig. 3. Major pathway for tyrosine catabolism. Major tyrosine metabolities include *p*-tyramine, *p*-hydroxyphenylpyruvic acid (*PHPAA*) *p*-hydroxyphenylacetic acid (*PHPAA*) *p*-hydroxyphenyllactic acid, (*PHPLA*) and homogentisic acid. Enzymes include tyrosine aminotransferase (*A*), *p*-hydroxphenylpyruvate oxidase hydrolyase (*B*), homogentisic and oxidase (*C*), and fumaryl acetoacetate (*FAA*) hydrolyase (*D*). (From Goldsmith and Laberge 1989, with permission.)

overloaded, and the unusual metabolites accumulate. These metabolites, succinylacetone in particular, are considered likely to produce toxic injury to the liver, resulting in liver failure as observed in this acute infantile presentation of hereditary tyrosinemia. In the kidney the same metabolites can lead to renal tubular injury, but renal compromise has been mild in relation to the severity of hepatic injury. Attempts to control toxicity of accreted metabolites has taken the form of dietary restriction of tyrosine (and phenylalanine, its metabolic precursor) to a level of intake just sufficient for growth, without overload of the obstructed pathway. On such a regimen of restriction, it appears that some patients can be controlled. Their levels of succinylacetone are diminished, and decompensation is reversed or prevented. Less acute presentations of hereditary tyrosinemia have been associated with residual FAH enzyme activity. Many of these patients, even on dietary management, ultimately develop acute hepatic decompensation, triggered by catabolic stresses like intercurrent infection, that overload the tyrosine pathway. Recently, an additional apparently frequent complication involving neurologic crisis has been described (MITCHELL et al. 1990). This manifests as the acute onset of peripheral neuropathy, with muscle pain and weakness, the latter of which can manifest as respiratory paralysis. These crises were in some cases lethal. They resemble neurologic crises previously observed in porphyria, and, not surprisingly, it has been shown that the enzyme delta-aminolevulinic acid dehydratase, an enzyme of the porphyrin biosynthetic pathway, is profoundly inhibited by the tyrosinemia metabolite succinylacetone.

For those hereditary tyrosinemia patients who do not suffer acute hepatic or neurologic decompensation, the usual outcome is development of hepatocellular carcinoma. This may result from carcinogenic action of the accumulated metabolites, or perhaps from chronic metabolic injury with its attendant chronic proliferative stimulus. Taking into account demise from either liver failure or from hepatocellular carcinoma, the overall rate of survival of hereditary tyrosinemia is 50% for 1 year and less than 10% for 5 years. Most of the 5-year survivors proceed to develop malignant hepatoma (GOLDSMITH and LABERGE 1989).

2.3 Hepatic Transplantation

The effect of hepatic transplantation in the case pesented here was dramatic. The clinical and biochemical signs of this disease literally "went away." However, it should be pointed out that there are both immediate and long-term risks to hepatic transplantation. The immediate risks are perioperative complications—bleeding, vascular or biliary obstruction, donor organ failure, and graft rejection. These may produce an immediate mortality rate ranging from 20% to 50%. Because only a handful of transplant surgeons have developed the skills to transplant infants, and because only a small number of infants have been treated, these numbers are uncertain. Beyond the immediate postoperative period, the long-term tradeoff of hepatic transplantation is lifelong immunosuppression. Here, the most significant complications are infection or tumor development. For the patient presented here, the choice was straightforward; face the risks of complication from hepatic transplantation or face virtually immediate death.

2.4 Genetics, Prenatal Diagnosis

Hereditary tyrosinemia is an autosomal recessive disorder. The FAH enzyme is a cytosolic protein that proves to be a dimer of two identical 43K subunits. Its RNA has been enriched by immunoprecipitation of hepatic polysomes (NICOLE et al.

1986), and a cloned cDNA should be forthcoming. Despite a lack of cloned probes, prenatal diagnosis of FAH deficiency can be carried out by detection of accumulation of the metabolite succinylacetone in amniotic fluid (PETTIT et al. 1984). This appears to be sensitive and specific, although some alleles of FAH deficiency might not lend to sufficient production during intrauterine life to enable diagnosis. Because hereditary tyrosinemia is an unusual condition, prenatal diagnosis would in general be performed only for couples that have already had an affected infant. However, the condition has a particularly high frequency in Quebec Province, affecting one of every 700 births, and in that region the condition is routinely screened at birth, using measurement of bloodspot tyrosine. Wherever a high tyrosine is observed, subsequent measurement of succinylacetone in urine has been able to distinguish hereditary tyrosinemia from benign conditions of tyrosine accumulation.

3 Rationale for Considering New Approaches to Therapy of Inherited Liver Diseases

Thus, faced with either of the foregoing severe single gene defects involving gene products expressed in the liver, novel therapies of enzyme, gene, or organ replacement are reasonable considerations. It could be argued that with prenatal diagnosis available for both of these conditions, couples need only go through these catastrophic illnesses once—subsequent affected pregnancies could be therapeutically aborted. However, it should be pointed out that exercising abortion options, involving often the midtrimester delivery of a dead fetus, is itself an extreme measure and, to many couples, completely unacceptable. Thus new therapeutic options are desirable. Hepatic transplantation as employed for our patient with hereditary tyrosinemia can be a successful enterprise, but it is not without long-term tradeoffs. Could gene transfer do better? This is entirely unknown because the feasibility of efficient gene transfer and long-term maintenance of expression of an introduced sequence is unknown. Yet the development of such an approach should be pursued, because it holds promise both for the definitive treatment of these diseases, and for a host of others that are less severe.

4 Novel Strategies for Treatment

4.1 Cell Therapy

One general approach to correction of hepatic defects is to supply either normal hepatocytes or "corrected" hepatocytes that could provide enough of the deficient function to alter the course of the clinical defect. Two studies with the

Gunn rat, deficient of the hepatic enzyme bilirubin glucuronyl transferase, and with the analbuminemic rat, suggest that this can be accomplished (MATAS et al. 1976; DEMETRIOU et al. 1986). In both of these studies, hepatocyte suspensions were prepared from syngeneic "donor" animals. In the study by MATAS et al., cells were injected into the portal vein of recipient homozygous Gunn rats. Following such injection, the serum bilirubin levels of the recipients were lowered and in several cases remained so for as long as 30 weeks. Analysis of the recipient livers revealed that the injected cells had become localized to the sinusoids. In the study by DEMETRIOU et al., donor cells were attached to carrier beads, and the beads were injected into the peritoneal cavity of recipient animals. Transient correction of both glucuronyl transferase and analbuminemia defects were observed. In the case of the latter study, it is unknown whether such cells in an ectopic location, the peritoneal surface, could maintain long-term differentiated function, necessary, for example, to maintain expression of OTC or FAH. For both approaches, it is unclear whether a sufficient amount of cells could be delivered to effect a correction of urea cycle defects or FAH. A volume of hepatocytes totaling approximately 5% of the entire liver might be predicted to effect 10% normal activity (assuming hepatic-specific functions are maintained at 100% in the transferred cells). Finally, for both strategies, there is the issue of the immune system in the case of human transfer. The problems of graft tolerance might be circumvented by excising a portion of the patient's *own* liver to provide donor cells, and infecting these cells with retroviruses carrying the desired function. Such a strategy has in fact been tested. WILSON et al. (1990) infected primary cultures of hepatocytes prepared from the LDL receptor-deficient Watanabe rabbit with a retrovirus bearing an LDL receptor cDNA. The infected cell population, expressing LDL receptors, was inserted into the peritoneal cavity of an affected rabbit, and shortly thereafter a fall in serum cholesterol was observed. For all of the foregoing strategies the major problems to be solved involve maintaining long-term hepatocyte engraftment and maintenance of differentiated function.

4.2 DNA-Mediated Therapy

The possibility that genes could be targeted to the liver via the asialoglycoprotein receptor, which resides at the surface of hepatocytes, avidly binds desialated proteins, and subsequently directs them into the system of endosomal uptake, has recently been pursued. WU et al. (1989) were able to prepare a complex between desialylated orosomucoid coupled with polylysine and a plasmid DNA joining an albumin promoter with chloramphenicol acetyltransferase (CAT), and injected this complex intravenously into young rats. They found that CAT activity could be detected in liver extracts of these animals during the first 24 h after injection, but it was not maintained much beyond this time. However, if partial hepatectomy was carried out 30 min after the injection, persistent and high levels of activity could then be observed for at least 11 weeks after injection. The state of the inserted DNA was examined, and at least a portion was found to be integrated. This approach seems promising because it potentially avoids the constraint of viral packaging systems on size of transferred DNA. Entire genomic segments might be transferrable using this technique.

4.3 Virus Therapy

4.3.1 Retroviruses

The efficiency with which retroviruses infect dividing cell populations and the stability of the integrated proviral genome are very attractive properties for the pursuit of therapeutic genetic transduction. While these viruses may ultimately be utilized as agents of delivery to bone marrow cell populations and to epithelial cells, their utility for infecting hepatocytes may be limited by the fact that the liver is in general a nondividing organ. The requirement for a dividing cell population as a target for retrovirus infection is met only during development and following injury. Whether the strategy of infecting a developing liver could be employed, given the current ability to make very early prenatal diagnosis (for example, using chorionic villus sampling), is unknown. Clearly, the approach of taking cells out of the liver, placing them in culture, and infecting them as they divide has been shown to be feasible (LEDLEY et al. 1987; WOLFF et al. 1987; WILSON et al. 1988), although the problems of dedifferentiation and of transfer of sufficient cell numbers, as discussed above, remain to be solved. Thus, at present the potential utility of retroviruses for transducing genes to the liver requires further evaluation.

4.3.2 Hepadnaviruses

The family of hepatitis B viruses offers another potential means of delivering genes to the liver. Here, the problems of mitotic inactivity of the liver are obviated—the organ in its nondividing state is the normal target for these viruses. Further, given the large percentage of naturally infected human beings who continue to harbor the viral genome and express surface antigen from it, in many instances without ill effect (i.e., are chronic SAg carriers), it seems conceivable that long-term maintenance and expression of an infecting recombinant genome bearing a desired gene could be achieved.

Initial attempts to produce defective and recombinant hepadnavirus particles in cultured cells have been successful (HORWICH et al. 1990). These studies were based on the notion that precise removal of the 12 base direct repeat, DR1, from the duck hepatitis B viral genome would render that genome defective for reverse transcription into an infectious DNA genome, but would not interfere with its expression of the functions needed in *trans* to package and reverse transcribe a second, defective, genome. Thus if all that is required in *cis* in a defective genome is an RNA packaging signal, plus DR1, and DR2, then a large portion of a second genome, totaling greater than 2 kb, might be dispensable and could be replaced with a foreign cDNA sequence, for example. Indeed, consistent with this possibility, the pregenomic RNA packaging signal has recently been found to lie in the precore region, near DR1 (NIEPMANN and SCHALLER 1989). However, additional sequences may be required in *cis*. For examle, a recent study suggests that at least a portion of the *pol* gene is required in *cis* for efficient reverse transcription (HIRSCH et al. 1990).

It appears that defective hepadnaviral particles produced from a tissue culture system are capable of infecting hepatocytes. Summers and coworkers were able to harvest culture medium from HuH7 cells cotransfected with a DR1-deleted duck HBV dimer genome and a surface antigen-minus (polymerase plus) mutant dimer genome and used the culture fluid to infect a culture of primary duck hepatocytes (HORWICH et al. 1990; SUMMERS et al. 1990). After 10 days the hepatocytes were found to be replicating the infecting defective genome and to be expressing core but not surface antigen, as predicted by the genotype of the infecting genome. Thus transduction is possible. Whether recombinant particles can now be produced in sufficient titer to infect a substantial portion of a recipient liver remains to be seen. If that could be accomplished, the major unresolved question would concern the long-term maintenance and expression of an introduced genome. It seems likely that in a clinical setting infection of newborn animals would be desirable because, at least in humans, the carrier state is much more likely to be established with neonatal infection. Clearly, a great deal remains to be addressed before hepadnaviruses could be seriously considered as therapeutic vectors. However, using ducks and duck hepatitis B virus as an experimental system, it should be possible to thoroughly investigate the various questions prior to any considerations about human beings. Certainly, in addition to retroviruses and hepadnaviruses, other viruses such as herpesviruses and adenovirus may also be amenable to manipulation as therapeutic transducing agents. These viruses can also be tested in animal systems.

5 Less Severe Hepatic Disorders

If a safe and efficient system of cell or gene transfer to liver can be worked out, then a number of additional inherited single gene defects could potentially be corrected (see SCRIVER 1989 for additional details).

Phenylketonuria. Phenylketonuria (PKU) is a relatively common inborn error of metabolism affecting approximately one of every 10000 births. Deficiency of the hepatic enzyme phenylalanine hydroxylase results in defective conversion of phenylalanine to tyrosine, and, as a result, phenylalanine and accumulated metabolites produce toxic effects on the brain. This results in profound mental retardation. Statewide screening programs that test blood spots from all newborns for accumulation of phenylalanine have enabled rapid diagnosis of the condition in the newborn period and effective treatment by restriction of dietary phenylalanine. This has permitted normal development, but the diet is fairly restrictive and in many social settings compliance is poor. This most immediately results in loss of IQ points of the affected child. Thus a definitive therapy would be desirable.

Alpha-1-Antitrypsin Deficiency. In patients with deficiency of this enzyme, inadequate antiprotease activity results in lung disease produced by unchecked activity of elastase and other proteases. Typically such disease arises in middle adult life, particularly in individuals who smoke. Because this enzyme is normally produced by the liver and released into the circulation, it seems that it could just as easily be produced by cell or gene transfer into a nonhepatic site.

Carbohydrate Disorders. Inherited conditions, including several of the glycogen storage diseases, galactosemia, and hereditary fructose intolerance, involve hepatic enzymes and could be corrected by provision of the normal gene to hepatocytes. In the first condition, considering the most common defect, glucose-6-phosphatase deficiency (von Gierke's disease), this would lead to normal glycogen metabolism and would both remove the present need for frequent glucose or corn starch feedings and remove risks of catastrophic hypoglycemia. In the second disorder, it could remove the toxic effects of galactose metabolites that can damage the lens, brain, and ovary. In the third case, it would remove the risk of hepatic or systemic injury from toxic metabolites that accumulate following ingestion of fructose.

Disorders of Amino Acid Metabolism. In addition to the defects described earlier, OTC deficiency, CPS deficiency, and FAH deficiency, other disorders might also be corrected. Propionic acidemia and methylmalonic acidemia are disorders, often severe, that produce acidemia due to accumulation of unusual organic acids behind blocks in the pathway of three carbon metabolism. Similarly, the defect of branch chain amino acid degradation, maple syrup urine disease, might similarly be corrected.

Lipid Metabolism. Familial hypercholesterolemia, due to LDL receptor deficiency, might be a prime candidate as one of the most common dominant mutations of humans—1:500 individuals is a heterozygote. Ability to produce additional normal receptor at the hepatocyte surface would most likely lead to a lowering of plasma LDL in these individuals (YOKODE et al. 1990). Other types of defect of lipoprotein metabolism might also be corrected at the liver.

Porphyrin Metabolism. A major class of porphyrias involves the liver. For example, acute intermittent porphyria is the most common such defect. Ability to supply additional porphobilinogen deaminase activity might prevent the abdominal attacks and neurologic disturbances that characterize this disorder.

In summary, there is great therapeutic potential for a strategy that is genuinely safe and efficient. We have entered an era where animals models for many candidate disorders are being produced. These systems can be used to test and further develop the creative approaches that may emerge for this class of disorders.

Acknowledgement. I thank Wayne Fenton and Saul Brusilow for helpful discussions and encouragement during preparation of this manuscript.

References

Arne PH, Houser ER, Thomas GH, Herman G, Hess D, Brusilow S (1990) Episodes of hyperammonemic coma, often postpartum, occurring in women who have a mutation at the OTC locus: New Engl J Med 322: 1652–1655
Batshaw ML, Brusilow S, Waber L, Blom W, Brubakk AM, Burton BK, Can HM, Kerr D, Mamunes P, Matalon R, Myerberg D, Schafer IA (1982) Treatment of inborn errors of urea synthesis: activation of alternative pathways of waste nitrogen synthesis and excretion. N Engl J Med 306: 1387–1392
Brusilow SW, Horwich AL (1989) Urea cycle enzymes. In Scriver CL, Beaudet AL, Sly WS, Valle D (eds) The metabolic basis of inherited disease, 6th edn. McGraw-Hill, New York, pp 629–663
Brusilow SW, Tinker J, Batshaw ML (1980) Amino acid acylation a mechanism of nitrogen excretion in inborn errors of urea synthesis. Science 207: 659–661
Brusilow SW, Danney M, Waber LJ, Batshaw M, Burton B, Levitsky L, Roth K, McKeethren C, Ward J (1984) Treatment of episodic hyperammonemia in children with inborn errors of urea synthesis. N Engl J Med 310: 1630–1634
Demetriou AA, Whiting JF, Feldman D, Levenson SM, Chowdhury NR, Moscion AD, Kram M, Chowdhury JR (1986) Replacement of liver function in rats by transplantation of microcarrier-attached hepatocytes. Science 233: 1190–1192
Fox JE, Rosenberg LE (1988) Toward a molecular understanding of ornithine transcarbamylase deficiency. Adv Neurol 48: 71–81
Fox JE, Hack AM, Fenton WA, Golbus MS, Winter S, Kalousek F, Rozen R, Brusilow SW, Rosenberg LE (1986) Prenatal diagnosis of ornithine transcarbamylase deficiency with use of DNA polymorphisms. N Engl J Med 315: 1205–1208
Grompe M, Muzny DM, Caskey CT (1989) Scanning detection of mutations in human ornithine transcarbamylase by chemical mismatch cleavage. Proc Natl Acad Sci USA 86: 5888–5892
Goldsmith LA, LaBerge C (1989) Tyrosinemia and related disorders. In Scriver CL, Beaudet AL, Sly WS, Valle D (eds) The metabolic basis of inherited disease, 6th edn McGraw-Hill, New York, pp 556–562
Haldane JBS (1935) The rate of spontaneous mutation of a human gene. J Genet 31: 317–326
Hata A, Tsuzuki T, Shimada K, Takiguchi M, Mori M, Matsuda I (1988) Structure of the human ornithine transcarbamylase gene. J Biochem (Tokyo) 103: 302–308
Hata A, Setoyama C, Shimada K, Takeda E, Koruda Y, Akaboshi I, Matsuda I (1989) Ornithine transcarbamylase deficiency resulting from a C to T substitution in exon 5 of the ornithine transcarbamylase gene. Am J Hum Genet 45: 123–127
Hirsch RC, Lavine JE, Chang LJ, Varmus HE, Ganem (1990) Polymerase gene products of hepatitis B viruses are required for genomic RNA packaging as well as for reverse transcription. Nature 344: 552–555
Hodges PE, Rosenberg LE (1989) The spf-ash mouse a missense mutation in the ornithine transcarbamylase gene also causes aberrant mRNA splicing. Proc Natl Acad Sci USA 86: 4142–4146
Holzgreve W, Golbus MS (1984) Prenatal diagnosis of ornithine transcarbamylase deficiency utilizing fetal liver biopsy. Am J Hum Genet 36: 320–328
Horwich AL, Fenton WA, Williams KR, Kalousek F, Kraus JP, Doolittle RF, Konigsberg W, Rosenberg

LE (1984) Structure and expression of a complementary DNA for the nuclear coded precursor of human mitochondrial ornithine transcarbamylase. Science 224: 1068–1074
Horwich AL, Furtak K, Pugh J, Summers J (1990) Synthesis of hepadnavirus particles that contain replication-defective duck hepatitis B virus genomes in cultured HuH7 cells. J Virol 64: 642–650
Houser ER, Finkelstein JE, Valle D, Brusilow S (1990) Allopurinol-induced orotidinuria a test for mutations at the ornithine transcarbamylase locus in women. N Engl J Med 322: 1641–1645.
Ledley FD, Darlington GJ, Tahn T, Woo SLC (1987) Retroviral gene transfer into primary hepatocytes: implications for genetic therapy of liver-specific functions. Proc Natl Acad Sci USA 84: 5335–5339
Lindgren V, deMartinville B, Horwich AL, Rosenberg LE, Francke U (1984) Human ornithine transcarbamylase locus mapped to band Xp21.1 near the Duchenne muscular dystrophy locus. Science 226: 698–700
Lyon MF (1961) Gene action in the X-chromosome of the mouse (mus musculus). Nature 190: 372–373
Maddalena A, Spence JE, O'Brien WE, Nussbaum RL (1988) Characterization of point mutations in the same arginine codon in three unrelated patients with ornithine transcarbamylase deficiency. J Clin Invest 82: 1353–1358
Matas AJ, Sutherland DER, Steffes MW, Mauer SM, Lowe A, Simmons L, Najarian JS (1976) Hepatocellular transplantation for metabolic deficiencies: decrease of plasma bilirubin in Gunn rats. Science 192: 892–894
Mitchell G, LaRochelle J, Lambert M, Michaud J, Grenier A, Ogier, Gauthier M, LaCroix J, Vanasse M, Larbrisseau A, Paradis K, Weber A, Lefevr Y, Melancon S, Dallaire L (1990) Neurologic crises in hereditary tyrosinemia. N Engl J Med 322: 432–437
Nicole LM, Valet JP, LaBerge C, Tanguay RM (1986) Purification of mRNA coding for the enzyme deficient in hereditary tyrosinemia, fumarylacetoacetate hydrolase. Biochem Cell 64: 489–
Niepmann M, Schaller H (1989) The precore/core region of hepatitis B virus RNA pregenome contains the signal for pregenome encapsidation. Hepatitis B viruses abstracts. Cold Spring Harbor Laboratory, Cold Spring Harbor, p 11
Petti BR, MacKenzie F, King GS (1984) The antenatal diagnosis and aid to the management of hereditary tyrosinemia by use of a specific and sensitive GC-MS assay for succinylacetone. J. inherited Metab Dis [Suppl] 2: 135–136
Rozen R, Fox J, Fenton WA, Horwich AL, Rosenberg LE (1985) Gene deletion and restriction fragment polymorphisms at the human ornithine transcarbamylase locus. Nature 313: 815–817
Rutledge SL, Havens PL, Haymond MW, McLean RH, Kan JS, Brusilow SW (1990) Neonatal hemodialysis: effective therapy for the encephalopathy of inborn errors of metabolism. J Pediatr 116: 125–128
Scriver CR et al (1989) The metabolic basis of inherited disease, 6th edn. McGraw-Hill, New York
Spence JE, Maddalena A, O'Brien WE, Fernbach SD, Batshaw ML, Leonard CO, Beaudet AL (1989) Prenatal diagnosis and heterozygote detection by DNA analysis in ornithine transcarbamylase deficiency. J Pediatr 114: 582–588
Summers J, Smith PM, Horwich AL (1990) Hepadnavirus envelope proteins regulate covalently closed circular DNA amplification. J Virol 64: 2819–2824
Veres G, Gibbs RA, Scherer SE, Caskey CT (1987) The molecular basis of the sparse fur mouse mutation. Science 237: 415–417
Wilson JM, Jefferson DM, Choudhury J, Novikoff PM, Johnston DE, Mulligan RC (1988) Retrovirus mediated transduction of adult hepatocytes. Proc Natl Acad Sci USA 85: 3014–3018
Wilson JM, Choudhury NR, Grossman M, Wajsman R, Epstern A, Mulligan RC, Choudhury JR Temporary amelioration of hyperlipidemia in low density lipoprotein receptor-deficient rabbits transplanted with genetically modified hepatocytes. Proc Natl Acad Sci USA 87: 8437–8441
Wolff JA, Yee J-K, Skelly HF, Moores JC, Respess JG, Friedman T, Leffert H (1987) Expression of retrovirally transduced genes in primary cultures of adult rat hepatocytes. Proc Natl Acad Sci USA 84: 3344–3348
Wu CH, Wilson JM, Wu GY (1989) Targeting genes: delivery and persistent expression of a foreign gene driven by mammalian regulatory elements *in vivo*. J Biol Chem 264: 16985–16987
Yokode M, Hammer RE, Ishibashi S, Brown MS, Goldstein JL (1990) Science 250: 1273–1275

Subject Index

Current Topics in Microbiology and Immunology

Volumes published since 1985 (and still available)

Vol. 116: **Willis, Dawn B. (Ed.):** Iridoviridae. 1985. 65 figs. X, 173 pp. ISBN 3-540-15172-9

Vol. 122: **Potter, Michael (Ed.):** The BALB/c Mouse. Genetics and Immunology. 1985. 85 figs. XVI, 254 pp. ISBN 3-540-15834-0

Vol. 124: **Briles, David E. (Ed.):** Genetic Control of the Susceptibility to Bacterial Infection. 1986. 19 figs. XII, 175 pp. ISBN 3-540-16238-0

Vol. 125: **Wu, Henry C.; Tai, Phang C. (Ed.):** Protein Secretion and Export in Bacteria. 1986. 34 figs. X, 211 pp. ISBN 3-540-16593-2

Vol. 126: **Fleischer, Bernhard; Reimann, Jörg; Wagner, Hermann (Ed.):** Specificity and Function of Clonally Developing T-Cells. 1986. 60 figs. XV, 316 pp. ISBN 3-540-16501-0

Vol. 127: **Potter, Michael; Nadeau, Joseph H.; Cancro, Michael P. (Ed.):** The Wild Mouse in Immunology. 1986. 119 figs. XVI, 395 pp. ISBN 3-540-16657-2

Vol. 128: 1986. 12 figs. VII, 122 pp. ISBN 3-540-16621-1

Vol. 129: 1986. 43 figs., VII, 215 pp. ISBN 3-540-16834-6

Vol. 130: **Koprowski, Hilary; Melchers, Fritz (Ed.):** Peptides as Immunogens. 1986. 21 figs. X, 86 pp. ISBN 3-540-16892-3

Vol. 131: **Doerfler, Walter; Böhm, Petra (Ed.):** The Molecular Biology of Baculoviruses. 1986. 44 figs. VIII, 169 pp. ISBN 3-540-17073-1

Vol. 132: **Melchers, Fritz; Potter, Michael (Ed.):** Mechanisms in B-Cell Neoplasia. Workshop at the National Cancer, Institute, National Institutes of Health, Bethesda, MD, USA, March 24–26, 1986. 1986. 156 figs. XII, 374 pp. ISBN 3-540-17048-0

Vol. 133: **Oldstone, Michael B. (Ed.):** Arenaviruses. Genes, Proteins, and Expression. 1987. 39 figs. VII, 116 pp. ISBN 3-540-17246-7

Vol. 134: **Oldstone, Michael B. (Ed.):** Arenaviruses. Biology and Immunotherapy. 1987. 33 figs. VII, 242 pp. ISBN 3-540-17322-6

Vol. 135: **Paige, Christopher J.; Gisler, Roland H. (Ed.):** Differentiation of B Lymphocytes. 1987. 25 figs. IX, 150 pp. ISBN 3-540-17470-2

Vol. 136: **Hobom, Gerd; Rott, Rudolf (Ed.):** The Molecular Biology of Bacterial Virus Systems. 1988. 20 figs. VII, 90 pp. ISBN 3-540-18513-5

Vol. 137: **Mock, Beverly; Potter, Michael (Ed.):** Genetics of Immunological Diseases. 1988. 88 figs. XI, 335 pp. ISBN 3-540-19253-0

Vol. 138: **Goebel, Werner (Ed.):** Intracellular Bacteria. 1988. 18 figs. IX, 179 pp. ISBN 3-540-50001-4

Vol. 139: **Clarke, Adrienne E.; Wilson, Ian A. (Ed.):** Carbohydrate-Protein Interaction. 1988. 35 figs. IX, 152 pp. ISBN 3-540-19378-2

Vol. 140: **Podack, Eckhard R. (Ed.):** Cytotoxic Effector Mechanisms. 1989. 24 figs. VIII, 126 pp. ISBN 3-540-50057-X

Vol. 141: **Potter, Michael; Melchers, Fritz (Ed.):** Mechanisms in B-Cell Neoplasia 1988. Workshop at the National Cancer Institute, National Institutes of Health, Bethesda, MD, USA, March 23–25, 1988. 1988. 122 figs. XIV, 340 pp. ISBN 3-540-50212-2

Vol. 142: **Schüpach, Jörg:** Human Retrovirology. Facts and Concepts. 1989. 24 figs. 115 pp. ISBN 3-540-50455-9

Vol. 143: **Haase, Ashley T.; Oldstone Michael B. A. (Ed.):** In Situ Hybridization 1989. 22 figs. XII, 90 pp. ISBN 3-540-50761-2

Vol. 144: **Knippers, Rolf; Levine, A. J. (Ed.):** Transforming Proteins of DNA Tumor Viruses. 1989. 85 figs. XIV, 300 pp. ISBN 3-540-50909-7

Vol. 145: **Oldstone, Michael B. A. (Ed.):** Molecular Mimicry. Cross-Reactivity between Microbes and Host Proteins as a Cause of Autoimmunity. 1989. 28 figs. VII, 141 pp. ISBN 3-540-50929-1

Vol. 146: **Mestecky, Jiri; McGhee, Jerry (Ed.):** New Strategies for Oral Immunization. International Symposium at the University of Alabama at Birmingham and Molecular Engineering Associates, Inc. Birmingham, AL, USA, March 21–22, 1988. 1989. 22 figs. IX, 237 pp. ISBN 3-540-50841-4

Vol. 147: **Vogt, Peter K. (Ed.):** Oncogenes. Selected Reviews. 1989. 8 figs. VII, 172 pp. ISBN 3-540-51050-8

Vol. 148: **Vogt, Peter K. (Ed.):** Oncogenes and Retroviruses. Selected Reviews. 1989. XII, 134 pp. ISBN 3-540-51051-6

Vol. 149: **Shen-Ong, Grace L. C.; Potter, Michael; Copeland, Neal G. (Ed.):** Mechanisms in Myeloid Tumorigenesis. Workshop at the National Cancer Institute, National Institutes of Health, Bethesda, MD, USA, March 22, 1988. 1989. 42 figs. X, 172 pp. ISBN 3-540-50968-2

Vol. 150: **Jann, Klaus; Jann, Barbara (Ed.):** Bacterial Capsules. 1989. 33 figs. XII, 176 pp. ISBN 3-540-51049-4

Vol. 151: **Jann, Klaus; Jann, Barbara (Ed.):** Bacterial Adhesins. 1990. 23 figs. XII, 192 pp. ISBN 3-540-51052-4

Vol. 152: **Bosma, Melvin J.; Phillips, Robert A.; Schuler, Walter (Ed.):** The Scid Mouse. Characterization and Potential Uses. EMBO Workshop held at the Basel Institute for Immunology, Basel, Switzerland, February 20–22, 1989. 1989. 72 figs. XII, 263 pp. ISBN 3-540-51512-7

Vol. 153: **Lambris, John D. (Ed.):** The Third Component of Complement. Chemistry and Biology. 1989. 38 figs. X, 251 pp. ISBN 3-540-51513-5

Vol. 154: **McDougall, James K. (Ed.):** Cytomegaloviruses. 1990. 58 figs. IX, 286 pp. ISBN 3-540-51514-3

Vol. 155: **Kaufmann, Stefan H. E. (Ed.):** T-Cell Paradigms in Parasitic and Bacterial Infections. 1990. 24 figs. IX, 162 pp. ISBN 3-540-51515-1

Vol. 156: **Dyrberg, Thomas (Ed.):** The Role of Viruses and the Immune System in Diabetes Mellitus. 1990. 15 figs. XI, 142 pp. ISBN 3-540-51918-1

Vol. 157: **Swanstrom, Ronald; Vogt, Peter K. (Ed.):** Retroviruses. Strategies of Replication. 1990. 40 figs. XII, 260 pp. ISBN 3-540-51895-9

Vol. 158: **Muzyczka, Nicholas (Ed.):** Viral Expression Vectors. 1991. Approx. 20 figs. Approx. XII, 190 pp. ISBN 3-540-52431-2

Vol. 159: **Gray, David; Sprent, Jonathan (Ed.):** Immunological Memory. 1990. 38 figs. XII, 156 pp. ISBN 3-540-51921-1

Vol. 160: **Oldstone, Michael B. A.; Koprowski, Hilary (Ed.):** Retrovirus Infections of the Nervous System. 1990. 16 figs. XII, 176 pp. ISBN 3-540-51939-4

Vol. 161: **Racaniello, Vincent R. (Ed.):** Picornaviruses. 1990. 12 figs. X, 194 pp. ISBN 3-540-52429-0

Vol. 162: **Roy, Polly; Gorman, Barry M. (Ed.):** Bluetongue Viruses. 1990. 37 figs. X, 200 pp. ISBN 3-540-51922-X

Vol. 163: **Turner, Peter C.; Moyer, Richard W. (Ed.):** Poxviruses. 1990. 23 figs. X, 210 pp. ISBN 3-540-52430-4

Vol. 164: **Bækkeskov, Steinnun; Hansen, Bruno (Ed.):** Human Diabetes. 1990. 9 figs. X, 198 pp. ISBN 3-540-52652-8

Vol. 165: **Bothwell, Mark (Ed.):** Neuronal Growth Factors. 1991. 14 figs. IX, 173 pp. ISBN 3-540-52654-4

Vol. 166: **Potter, Michael; Melchers, Fritz (Ed.):** Mechanisms in B-Cell Neoplasia 1990. 143 figs. XIX, 380 pp. ISBN 3-540-52886-5

Vol. 167: **Kaufmann, Stefan H. E. (Ed.):** Heat Shock Proteins and Immune Response. 1991. 18 figs. IX, 214 pp. ISBN 3-540-52857-1

MIX
Papier aus verantwortungsvollen Quellen
Paper from responsible sources
FSC® C105338

If you have any concerns about our products,
you can contact us on
ProductSafety@springernature.com

In case Publisher is established outside the EU,
the EU authorized representative is:
Springer Nature Customer Service Center GmbH
Europaplatz 3, 69115 Heidelberg, Germany

Printed by Libri Plureos GmbH
in Hamburg, Germany